ATSDR

Public Health Assessment
Guidance Manual

LEWIS PUBLISHERS
Boca Raton   Ann Arbor   London   Tokyo

Library of Congress Cataloging-in-Publication Data

United States. Agency for Toxic Substances and Disease Registry.
    ATSDR public health assessment guidance manual.
        p.   cm.
    Includes bibliographical references.
    ISBN 0-87371-857-7
        1. Health risk assessment--Handbooks, manuals, etc.
    2. Environmental health--Handbooks, manuals, etc. I. Title.
    RA566.27.U55   1992
    363.72'872--dc20   92-24923
                                                                                            CIP

PRINTED IN THE UNITED STATES OF AMERICA
    4   5   6   7   8   9   0

Printed on acid-free paper.

FOREWORD

The Agency for Toxic Substances and Disease Registry's highest priority is the protection of public health. We address that priority through actions that mitigate or prevent adverse health effects and diminished quality of life resulting from exposures to hazardous substances in the environment. Because large numbers of hazardous waste sites and facilities can potentially affect public health, ATSDR must have an instrument of triage to determine where, and for whom, public health actions should be undertaken. The Agency's instrument of choice is the public health assessment, which characterizes the nature and extent of hazards and identifies communities where public health actions are needed. This Public Health Assessment Guidance Manual provides the environmental health professional with directions for implementing that important public health tool.

The Public Health Assessment Guidance Manual is the result of the combined efforts of ATSDR, Oak Ridge National Laboratory, and state health departments participating in the ATSDR Public Health Assessment Cooperative Agreement Program. The draft manual was made available for public comment through an announcement in the Federal Register and distributed to federal, state, and local entities, private consultants and corporations, and trade/professional organizations. Comments received were considered and, when appropriate, incorporated into the manual. ATSDR is responsible for the manual's technical accuracy and its presentation of environmental health science practice.

The Public Health Assessment Guidance Manual is not intended to supplant the professional judgment and discretion of the health assessor (or public health assessment team) in compiling and analyzing data, drawing conclusions, and making public health recommendations. Instead, the manual provides a logical approach to evaluating the public health implications of hazardous waste sites, while still allowing health assessors to develop new approaches to the process and apply the most current and appropriate science and methodology. That is an important concept; just as environmental health science is rapidly developing, the public health assessment must also adapt to changing scientific technology and procedures in order to remain a dynamic process.

ATSDR is committed to updating the Public Health Assessment Guidance Manual as new technical information becomes available. The Agency welcomes comments from users of the manual.

Robert C. Williams, P.E.
Director, Division of Health
Assessment and Consultation

Barry L. Johnson, Ph.D.
Assistant Surgeon General
Assistant Administrator, ATSDR

The 1986 Superfund Amendments and Reauthorization Act (SARA) to the Comprehensive Environmental Response, Compensation, and Liability Act of 1980 (CERCLA) directs the Agency for Toxic Substances and Disease Registry (ATSDR) to perform specific public health activities associated with actual or potential exposure to hazardous substances released into the environment. Among these activities, ATSDR was mandated to perform a health assessment by December 10, 1988, for each facility listed or proposed to be listed on the National Priorities List (NPL). ATSDR must conduct health assessments on all sites added to the NPL since October 17, 1986, within one year of being listed (or proposed for listing). In addition, ATSDR may conduct a health assessment for a particular facility or release when petitioned by a person or group of persons.

U.S. DEPARTMENT OF HEALTH AND HUMAN SERVICES
PUBLIC HEALTH SERVICE
Agency for Toxic Substances and Disease Registry
Atlanta, Georgia 30333

TABLE OF CONTENTS

APPENDICES
    A - Environmental Media Evaluation Guides
    B - Standards and Health Guideline Values
    C - Data Evaluation Criteria
    D - Estimation of Exposure Dose
    E - Food Consumption Values
    F - State and Local Sources of Health Outcome Data
    G - National Sources of Health Outcome Data
    H - Health Outcome Data Checklist
    I - Sources of Health Outcome Data
    J - Census Information
    K - Units and Conversion Factors
    L - Site Summary Form
    M - Glossary

LIST OF FIGURES

LIST OF TABLES

AA	Atomic Absorption	FRDS	Federal Reporting Data System
ACGIH	American Conference of Governmental Industrial Hygienists	FS	Feasibility Study
		GC/MS	Gas Chromatography/Mass Spectrometry
AIC	Acceptable Intake for Chronic Exposures	HARP	Health Activities Recommendation Panel
AIS	Acceptable Intake for Subchronic Exposures	HAZDAT	Hazardous Substance Data Management Systetm
ATSDR	Agency for Toxic Substances and Disease Registry	HEAST	Health Effects Assessment Summary Tables
AWQC	Ambient Water Quality Criteria	HHAG	Human Health Assessment Group
BCF	Bioconcentration Factor		
CAA	Clean Air Act	HSDB	Hazardous Substances Data Bank
CERCLA	Comprehensive Environmental Response, Compensation, and Liability Act of 1980	HSL	Hazardous Substances List
		IARC	International Agency for Research on Cancer
CFR	Code of Federal Regulations		
CLP	Contract Laboratory Program	IRIS	Integrated Risk Information System
CRDL	Contract Required Detection Limits	Koc	Organic carbon partition coefficient (also known as soil/water partition coefficient)
CRQL	Contract Required Quantitation Limits		
CWA	Clean Water Act	Kow	Octanol-water partition coefficient
DHAC	Division of Health Assessment and Consultation (ATSDR)	LHA	Lifetime Health Advisory
DHS	Division of Health Studies (ATSDR)	LOAEL	Lowest Observed Adverse Effect Level
DWEL	Drinking Water Equivalent Level	MCL	Maximum Contaminant Level
EMEG	Environmental Media Evaluation Guide	MCLG	Maximum Contaminant Level Goal
EPA	Environmental Protection Agency	MRL	Minimal Risk Level
		MSA	Metropolitan Statistical Area
FDA	Food and Drug Administration		

NAAQS	National Ambient Air Quality Standards	QA/QC	Quality Assurance and Quality Control
NAS	National Academy of Sciences	RCRA	Resource Conservation and Recovery Act of 1976
NESHAP	National Emission Standards for Hazardous Air Pollutants	RfD	Reference Dose
NIOSH	National Institute for Occupational Safety and Health	RI	Remedial Investigation
		RI/FS	Remedial Investigation and Feasibility Study
NIPDWR	National Interim Primary Drinking Water Regulations	RPM	Remedial Project Manager (EPA)
NLM	National Library of Medicine		
NOAEL	No Observed Adverse Effect Level	SARA	Superfund Amendments and Reauthorization Act of 1986
NPDES	National Pollutant Discharge Elimination System	SDWA	Safe Drinking Water Act
		SOP	Standard Operating Procedures
NPDWR	National Primary Drinking Water Regulation	STEL	Short-Term-Exposure Limit
NPL	National Priorities List	STP	Standard Temperature and Pressure
NTP	National Toxicology Program	TAL	Target Analyte List
OERR	Office of Emergency and Remedial Response (EPA)	TCL	Target Compound List
		TIC	Tentatively Identified Compound
ORD	Office of Research and Development (EPA)	TLV	Threshold Limit Value
OSHA	Occupational Safety and Health Administration	TOXNET	Toxicology Data Network
		TRI	Toxic Chemical Release Inventory
OSWER	Office of Solid Waste and Emergency Response (EPA)	TWA	Time-Weighted Average
PEL	Permissible Exposure Limit	USGS	United States Geological Survey
PMCL	Proposed Maximum Contaminant Level	VOC	Volatile Organic Compound
PMCLG	Proposed Maximum Contaminant Level Goal	WQC	Water Quality Criteria
		XRF	X-Ray Fluorometry
PQL	Practical Quantitation Limit		
PWS	Public Water System		

1. INTRODUCTION

The Agency for Toxic Substances and Disease Registry (ATSDR) has three methods of conveying concern about a site's potential to cause adverse health effects: a public health advisory, a health assessment, and a health consultation.

The **ATSDR Public Health Advisory** is a communication from the ATSDR Administrator to the Administrator of the Environmental Protection Agency (EPA) that states ATSDR's concern that a public health threat exists of such importance and magnitude that immediate intervention should be taken by EPA. The health advisory is also provided to the appropriate EPA regional office and state health department.

The **ATSDR Public Health Assessment**, hereafter referred to in this guidance manual as the *health assessment*, is an analysis and statement of the public health implications posed by the facility or release under consideration. The health assessment is an evaluation of relevant environmental data, health outcome data, and community concerns associated with a site where hazardous substances have been released. The health assessment identifies populations living or working on or near hazardous waste sites for which more extensive public health actions or studies are indicated.

An **ATSDR Health Consultation** provides advice on specific public health issues that occur as a result of actual or potential human exposure to a hazardous material. The ATSDR health consultation, distinct from a health assessment, is a response to a question or request for information pertaining to a hazardous substance or facility, which includes waste sites. In addition, a health consultation often contains a time-critical element that necessitates a rapid response. A health consultation is, therefore, a more limited response than a health assessment.

This document sets forth the health assessment process as defined by ATSDR, clarifying the methodologies and guidelines that will be used by ATSDR staff and agents of ATSDR in conducting these health assessments. ATSDR may use selected extramural resources to help meet health assessment responsibilities and deadlines under the Comprehensive Environmental Response, Compensation, and Liability Act of 1980 (CERCLA) and the Resource Conservation and Recovery Act of 1976 (RCRA), as amended. Contractors may be used to support data management and information needs that are associated with a health assessment request. Cooperative agreements have been developed with some states to assist them in building their capacity to perform health assessments. This document is directed at ATSDR public health physicians, scientists and engineers, and other state and federal agencies that perform health assessments.

The individual steps for performing a health assessment are discussed in Chapters 3 through 8 of this document. Chapter 9 discusses how the health assessment report should be written and the format in which it should be presented.

The Agency for Toxic Substances and Disease Registry was created to implement the health-related sections of CERCLA 1980, as amended. One of the major vehicles for meeting that mandate is the health assessment. The following chapter provides the regulatory definition, purposes, and subsequent uses of a health assessment and contrasts the qualitative nature of a health assessment with the more quantitative risk assessment. The health assessment process is then described and the format for the health assessment is presented. These sections provide an introduction for the remainder of this manual, which is dedicated to providing guidance for conducting and writing health assessments.

2.1. DEFINITION
AND PURPOSE

A health assessment is the evaluation of data and information on the release of hazardous substances into the environment in order to assess any current or future impact on public health, develop health advisories or other recommendations, and identify studies or actions needed to evaluate and mitigate or prevent human health effects (55 Federal Register 5136, February 13, 1990, as codified at 42 Code of Federal Regulations Part 90).

As shown in Figure 2.1., the health assessment can be viewed as a platform founded on three sources of information:

Environmental characterization data for a hazardous waste site includes information on environmental contamination and environmental pathways. Such information is provided in site-specific Remedial Investigation (RI)

reports, preliminary assessments, Geological Surveys, and Site Inspection reports obtained from EPA and pertinent state and local environmental departments. A site visit, conducted by ATSDR staff, is also an important source of environmental characterization data.

Community health concerns associated with a site constitute a key data component of the health assessment. The community associated with a hazardous waste site includes the population living around the site, local public health officials, other local officials, and the local media. In order to acquire information on community health concerns, the health assessor must become an investigator; obtaining that information provides the health assessor with an opportunity to involve the public in the health assessment process. In addition, community health concerns can

Figure 2.1. Health Assessment Foundations

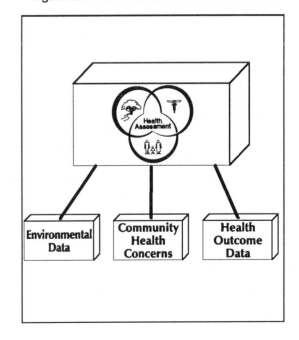

serve as a guide in evaluating health outcome data.

Health outcome data and parameters are the third major source of data for health assessments. The identification, review, and evaluation of health outcome parameters are interactive processes involving ATSDR, data source generators, and the community involved.

Health outcome data are community-specific and may include databases at the local, state, and national level, as well as data from private health care organizations and professional institutions and associations. Databases to be considered include medical records, morbidity and mortality data, tumor and disease registries, birth statistics, and surveillance data. Relevant health outcome data play an important role in assessing the public health implications associated with a hazardous waste site and in determining which follow-up health activities are needed.

The integration of environmental characterization data, community health concerns, and health outcome data is addressed in Chapters 3 through 8 of this manual.

The health assessment is a mechanism to respond to community health concerns associated with human exposure to hazardous substances at a site. As stated in the definition, a health assessment has three major purposes: 1) evaluating the public health implications of the site; 2) addressing those implications by developing health advisories or making recommendations, including further health or environmental studies; and 3) identifying populations where actions are necessary to mitigate or prevent adverse health effects.

When complete health or environmental data are lacking, ATSDR may determine it necessary to conduct further assessments for a site or facility as the data become available. A major reason for preparing a health assessment

is to determine the need for health effects studies at a site to further assess any current or future risks to public health. The health assessment is an evaluation of relevant environmental data, health outcome data, and community concerns associated with a site where hazardous substances have been released. The health assessment identifies populations living or working on or near hazardous waste sites for which public health actions are needed, such as health studies, health education, or chemical-specific research. ATSDR health assessments are based on factors such as the nature, concentration, toxicity, and extent of contamination at a site; the existence of potential pathways for human exposure; community health concerns; the size and nature of the community likely to be exposed; relevant community-specific, past and current, health outcome data; and any other information available to ATSDR that is relevant to a determination of potential risks to public health.

A health assessment is written for the informed community associated with the site, including citizen groups, local leaders, and health professionals. Health assessments are available for public review and comment, and their availability is announced in local communities. In addition, quarterly notices are placed in the *Federal Register* to announce health assessments completed in the previous three months. Health assessments are also intended to provide public health information to other government agencies (e.g., EPA and state health and environmental agencies). ATSDR supplies the EPA and state health agencies with a copy of each health assessment.

A health assessment for a facility or release is not always a single static document or report, but may be a series of reports over time that reflect the dynamic, iterative process of collecting and evaluating new information and data regarding the subject facility, site, or release. That iterative evaluation reflects assessments performed by ATSDR staff, and, under the Superfund Amendments and

Reauthorization Act (SARA) of 1986, it may reflect assessments performed by state or other local agencies. Further, SARA also directs that health assessments performed by such agencies be reported to the Administrators of both ATSDR and EPA and that recommendations resulting from such assessments be reflected in any reports or assessments issued by ATSDR. Thus, as more complete information is collected and evaluated, the conclusions and recommendations of the health assessment may be modified or altered to reflect the public health implications of the additional information.

Preliminary Health Assessments. ATSDR will designate as preliminary those health assessments prepared on sites for which site characterization is incomplete or for which no summary of relevant health outcome or environmental data exists. In those cases, ATSDR will attempt to summarize data and information from federal, state, and local (e.g., community) sources. Health assessments for those sites will be designated as preliminary to convey to EPA, the states, and the public that ATSDR has conducted the assessment based upon limited data. For those sites receiving a preliminary health assessment, a further health assessment will be performed, if considered appropriate, when the full environmental characterization is completed and provided to ATSDR for consideration.

Petitioned Health Assessments. Both CERCLA, as amended by SARA, and RCRA, as amended by the Hazardous Solid Waste Amendments of 1984, permit individuals and concerned parties to petition ATSDR to conduct health assessments. ATSDR has promulgated regulations describing the petitioned health assessment process, (42 Code of Federal Regulations, Part 90) published in 55 *Federal Register* 5136, February 13, 1990.

2.2. USES OF HEALTH ASSESSMENTS

As shown in Figure 2.2., the health assessment is central to many of the activities ATSDR performs. Health assessments are designed not only to evaluate health effects, but also to identify populations for which additional studies or public health actions are required. Thus, the health assessment is designed to identify:

1. knowledge gaps concerning the toxicity of substances identified at the facility or release under review;

2. communities near facilities or releases where biologic measurements of human exposure or medical investigations (e.g., community-based health outcome parameters) are needed; and

3. the need for additional health information (e.g., pilot studies, epidemiologic studies and registries, and site-specific surveillance).

ATSDR may then choose to initiate a variety of health studies based on the review of the health assessment, including: pilot health effects studies (disease- and symptom-prevalence studies, cluster investigations, exposure studies), epidemiologic studies, or disease registries. The method by which those studies are initiated and how they are related is shown in Figure 2.3.

As mentioned previously, a health assessment is written for the informed community associated with the site. Therefore, in addition to the uses identified above, the health assessment may be used by local, state and federal agencies, citizens, and other interested parties or individuals to identify the site's public health implications.

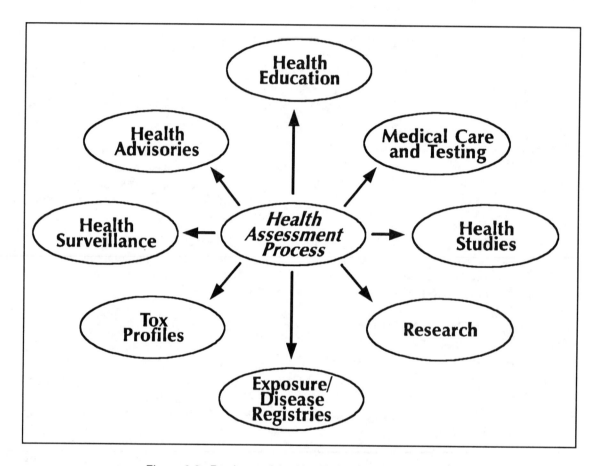

Figure 2.2. Products of the Health Assessment Process

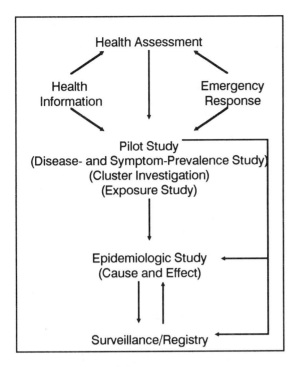

Figure 2.3. Relationship of Health Effects Studies

2.3. HEALTH ASSESSMENTS AND RISK ASSESSMENTS

Deliberate differences exist between ATSDR's health assessments and EPA's risk assessments. The two agencies have distinct purposes that necessitate different goals for their assessments.

Risk Assessments. A risk assessment is defined as a qualitative and quantitative process conducted by EPA to characterize the nature and magnitude of risks to public health from exposure to hazardous substances, pollutants, or contaminants released from specific sites. Risk assessments include the following components: hazard identification, dose-response assessment, exposure assessment, and risk characterization. Statistical and biologic models are used in quantitative and chemical-oriented risk assessments to calculate numeric estimates of

risk to health by using data from human epidemiologic investigations (when available) and animal toxicology studies. The product of quantitative risk assessment is a numeric estimate of the public health consequences of exposure to an agent. In preparing a risk assessment for a site, a risk assessor also attempts to include all adverse health effects, characterizing the risk to sensitive populations when the information is available. EPA risk assessments are used in risk management decisions to establish cleanup levels; to set permit levels for discharge, storage, or transport of hazardous waste; and to determine allowable levels of contamination.

Health Assessments. As discussed in Section 2.1., ATSDR health assessments are based on environmental characterization information, community health concerns, and health outcome data. Because of the nature of these databases, health assessments use quantitative as well as qualitative data, focusing on medical public health and toxicologic perspectives associated with exposure to a site. The health assessment specifically addresses community health concerns (e.g., sensitive populations, possible disease outcomes) and evaluates relevant, community-specific health outcome data. Combined with environmental data, information obtained from those two data sources are used to determine the public health implications of the site guiding the initiation of follow-up health activities when indicated.

Thus, while a risk assessment conducted under EPA's Remedial Investigation/Feasibility Study (RI/FS) process is used to support the selection of a remedial measure at a site, an ATSDR health assessment is a mechanism to provide the community with information on the public health implications of a specific site, identifying those populations for which further health actions or studies are needed. The health assessment also makes recommendations for actions needed to protect public health, which may include issuing health advisories.

2.4. FOLLOW-UP ACTIONS AND EVALUATIONS

ATSDR intends to conduct follow-up evaluations as necessary on all recommendations made in a health assessment. Under CERCLA Section 104(i)(6)(H), if the health assessment indicates that a release or threatened release may pose a serious threat to human health, the Administrator of ATSDR shall notify the Administrator of EPA who may place the site on the National Priorities List (NPL) or give the site a higher priority if it is already on the list. Further, the Administrator of ATSDR is empowered to conduct a pilot study of health effects, full-scale epidemiologic studies, disease registries, or surveillance studies in order to ensure that human health is protected. CERCLA Section 104(i)(11) states that if a health assessment finds that a significant risk to human health exists at a site, the President shall take steps to reduce and eliminate or substantially mitigate the threat to human health.

Each health assessment also contains a recommendation that describes follow-up health actions proposed for the site. Those recommendations are provided by the ATSDR Health Activities Recommendation Panel (HARP). In addition to making the recommendations, HARP is also responsible for tracking the recommendations and ensuring their implementation.

Finally, as a matter of policy and as situations warrant, ATSDR will contact the EPA and state and local agencies periodically to determine if land use, demographics, and other site conditions have changed to the extent that a follow-up health assessment is necessary. Addenda to the health assessment are written when necessary, based on information obtained in follow-up inquiries.

2.5. HEALTH ASSESSMENT PROCESS

To evaluate the public health implications posed by contamination at a site, the assessor must obtain and evaluate data and information on the site's history, the types and levels of contamination at the site, site-specific environmental transport mechanisms, routes of human exposure, community health concerns, relevant health outcome parameters, and medical and toxicologic implications of the site's contaminants. This evaluation is an iterative, dynamic process that considers available data from varying perspectives (Figure 2.4.). The interrelationship of the six steps will vary from site to site depending on the site's individual and unique characteristics.

Every health assessment includes six basic steps for acquiring the data and information necessary to evaluate the site's health risks:

1. **Evaluating information on the site's physical, geographical, historical, and operational setting;**

2. **Identifying health concerns of the affected community(ies);**

3. **Determining contaminants of concern associated with the site;**

4. **Identifying and evaluating exposure pathways (environmental transport mechanisms and human exposure pathways);**

5. **Determining public health implications based on available community-specific health outcome databases and other medical and toxicologic information; and**

6. **Determining conclusions and recommendations concerning the health threat posed by the site. The recommendations include the Public**

Figure 2.4. Factors Influencing the Health Assessment Process

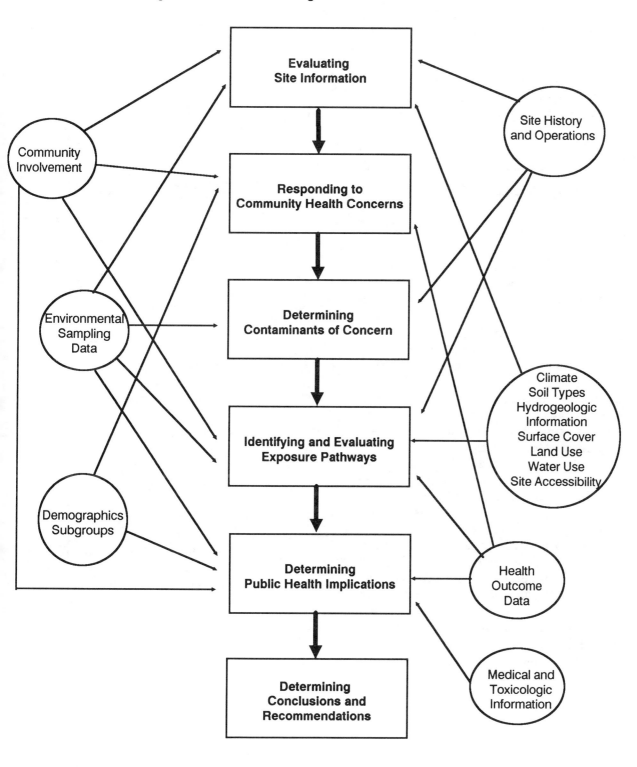

Health Actions, which specify the actions that have been taken or will be taken by ATSDR and other federal, state, and local agencies.

Information reviewed for each step in the health assessment process is evaluated for adequacy of data and potential health impacts at a hazardous waste site. Consideration is given to known past or expected future contamination and exposures.

2.6. HEALTH ASSESSMENT REPORT

The format shown in Table 2.1. was chosen to meet ATSDR's requirements for a health assessment (i.e., assess past, current, or future adverse public health effects; develop health advisories or other recommendations; and identify studies or actions needed to evaluate and mitigate or prevent human health effects). The first four sections of the health assessment are devoted to providing relevant background information, documenting community health concerns, identifying contaminants of concern and physical hazards, and evaluating environmental and human exposure pathways. The fifth section addresses the public health implications associated with the site based on toxicologic and relevant health outcome data evaluations. The final sections provide conclusions and recommendations for follow-up studies or mitigative or preventive actions. Specific follow-up health actions are presented under Public Health Actions. Guidance for determining and evaluating relevant information for each section of the health assessment is provided in Chapters 3 through 8 of this manual. Chapter 9 provides specific guidance for following the health assessment format when writing the report.

Table 2.1. Health Assessment Format

Summary

Background
A. Site Description and History
B. Site Visit
C. Demographics, Land Use, and Natural Resources Use
D. Health Outcome Data

Community Health Concerns

Environmental Contamination and Other Hazards
A. On-Site Contamination
B. Off-Site Contamination
C. Quality Assurance and Quality Control
D. Physical and Other Hazards

Pathways Analyses
A. Completed Exposure Pathways
B. Potential Exposure Pathways

Public Health Implications
A. Toxicologic Evaluation
B. Health Outcome Data Evaluation
C. Community Health Concerns Evaluation

Conclusions

Recommendations (Public Health Actions)

Preparers of Report

References

Appendices

3. EVALUATING SITE INFORMATION

No analysis of the threat posed by site contamination can begin without the health assessor first becoming familiar with the site, its setting, and its history. ATSDR considers a site visit to be an indispensable element of the health assessment process. The site visit allows the health assessor to determine current conditions at the site and to get firsthand knowledge of community health concerns. Site visits and regulatory reports provide much of the necessary site characterization information. Community interaction and personal contacts developed during the site visit are important means of obtaining relevant documents and gathering additional information. To begin a health assessment, the assessor needs to have site-specific knowledge about the following categories:

1. **Background** (site description, operations, history, regulatory involvement);

2. **Community health concerns;**

3. **Demographics** (characterization of populations at and near the site that may potentially be exposed to hazardous materials);

4. **Land use and natural resource use information** (land use and natural resources at and adjacent to the site);

5. **Environmental contamination** (chemicals and concentrations identified in specific media);

6. **Environmental pathways** (contaminant fate and transport mechanisms within the respective media); and

7. **Health outcome data.**

Specific information for each of those categories is presented in Table 3.1. Not all the information indicated in Table 3.1. is necessary to perform every health assessment; nor should

this table be considered inclusive for conducting a health assessment. However, in general, the greater the site-specific knowledge the assessor has about those seven categories of information, the greater the accuracy of health assessment conclusions.

The assessor must be familiar with the site information that is available and the usefulness of that information in conducting the health assessment. To the extent practicable, the assessor should investigate community health concerns and seek environmental and health outcome data. When little information is available about a site, the health assessment should be designated as "preliminary," indicating that a complete health assessment may be conducted when sufficient information is available. Information categories used in the health assessment will be discussed in the following sections of this chapter.

3.1. BACKGROUND INFORMATION

The assessor needs to have an understanding of the site, its history and operations, and its relation to the community surrounding it. This information will assist the assessor in understanding the potential nature, magnitude, and extent of contamination and the community health concerns related to site operations.

3.1.1. Site Description

Site descriptions provide background information that will be evaluated in greater detail during the health assessment. The assessor should determine the boundaries of the site in order to delineate the on- and off-site areas. The assessor should also note the

Table 3.1. Useful Data for Performing a Health Assessment

BACKGROUND INFORMATION

Site Description

☐ Site name(s), if any

☐ Site address or location

☐ Copy of U.S. Geological Survey (USGS) quadrangle map

☐ Site map showing distance from the site to closest existing residence (or where residence may be built)

☐ Political geography (incorporated cities or towns, counties, states)

☐ Site type (waste tailings, landfill, surface impoundment, spills)

☐ Number and types of National Priorities List (NPL), RCRA, and industrial sites within one-half mile of the site

☐ Information on hazardous materials releases, such as that found in the Toxic Chemical Release Inventory (TRI)

☐ Contact person (local, state, federal - name and phone number)

Site History

☐ Dates of operation, process description, and significant events

☐ Description of prior release and actions taken by EPA to remedy problems at the site

☐ Description of physical barriers to prevent pollutant transport (i.e., liners, slurry walls, fences, dikes)

☐ Current CERCLA or RCRA status of the site and site status with respect to remedial activities (past, current, and future)

COMMUNITY CONCERNS

☐ Records of environmental and health complaints by public about the site

☐ Health and other concerns gained through community meetings

☐ Log of actions taken by state or county health unit at or near the site in response to health issues, concerns, and complaints

HEALTH OUTCOME DATA

☐ Community health records and studies that may have been performed in the community

☐ Identity and source of relevant health outcome databases

DEMOGRAPHIC INFORMATION

☐ Approximate population potentially affected by the site

☐ Indicators of sensitive populations in the vicinity of the site (i.e., schools, nurseries, hospitals, retirement homes)

☐ Ethnic identity, age, and sex

LAND USE AND NATURAL RESOURCE USE INFORMATION

☐ Types of barriers or signs to prevent public access

☐ Activities on the site (and estimated number of people involved in each activity)

☐ Estimated frequency and types of on-site activities (e.g., dirt biking, camping, hunting, and fishing)

☐ Anticipated future land use or development

☐ Photographs that depict site conditions, quantity of waste, proximity to populated areas, and site use

☐ Map showing locations and uses of wells or springs within two miles of the site

☐ Approximate number of persons using groundwater for potable purposes

☐ Stream classifications and water uses downstream from the site

Table 3.1. Continued

☐ Agriculture, aquaculture, animal husbandry, hunting, and fishing near the site

ENVIRONMENTAL CONTAMINATION INFORMATION

Substances Present

☐ List of chemicals by descriptive name and Chemical Abstracts Service registry number

☐ Concentrations of contaminated materials present in different media (soil, air, water)

☐ Any chemical, mechanical, meteorologic, or other phenomena that may rapidly alter any of the following: current physical state and general condition of the substances and current structural condition of containers, vessels, and buildings that hold substances

☐ Summary of current and historical sampling data for all media

Quality Assurance and Quality Control

☐ Data Review Summary prepared by EPA Regional staff documenting the validation of sample holding times, instrument performance, calibration, blanks, surrogate recovery, matrix spike recovery, and compound identification. This includes documentation of actions taken to resolve data quality problems and an overall case assessment.

☐ For non-Contract Laboratory Program (CLP) data, equivalent information should be provided.

ENVIRONMENTAL PATHWAYS INFORMATION

Groundwater

☐ Geologic profile

☐ Map of water table contours and monitoring wells, if any

☐ Average net rainfall and evaporation rate

☐ Hydraulic conductivity of saturated zone (estimated or measured)

☐ Sampling data and description of sampling strategy with summary table, including location of contaminated monitoring wells and contaminant concentrations over time

Surface Water

☐ Map of 100-year floodplain

☐ Sampling data and description of sampling strategy with summary table

Soil and Sediment

☐ Sampling data and description of sampling strategy with summary table

☐ Soil type and characteristics

☐ Surface cover

Air

☐ Climatic information

☐ Wind rose

☐ Sampling data and description of sampling strategy with summary table

☐ Sampling data for subsurface gas migration

Biota

☐ Biologic sampling data (fish, animals, plants)

geographical features, land resource use, and demographic characteristics of the population surrounding the site.

Geographic location of the site provides insight about climatic and geologic conditions, floodplains, and location of major surface-water bodies.

Location of the site within the community provides insight about both the size of the population potentially affected by the site and other sources of contamination.

Visual representations of the site (site plans, topographic maps, and aerial photographs), in addition to identifying items mentioned in the previous paragraphs, indicate the size of site operations, extent of surface contamination, underground conduits for potential contaminant transport, distances to populations near the site, the presence of schools and hospitals, and land uses near the site.

Physical hazards at a site may constitute a public health concern. All aspects of a site's health implications are covered in a health assessment. Physical hazards that may be observed during the site visit include stacked drums, accessible chemical products, pits, dams, dikes, and unsafe structures (buildings, storage tanks).

3.1.2. Site History and Operations

Information on the site's historical development often provides an indication of the contaminants that may be present, the extent of contamination, the consequent rate of migration, and the magnitude of human exposure. Documents must be reviewed for the following information :

Types of activities carried out at the site will likely give an indication of the contaminants of concern at the site.

Duration of commercial and industrial activities at the site is likely to influence the extent of contamination and contaminant migration.

The length of time contamination has been present at the site may give an indication of the extent of contaminant migration and populations potentially exposed.

Changes in size or development of the site may drastically affect the rates and patterns of contaminant migration.

Current and planned remedial activities will affect the site's health implications and will need to be discussed in the health assessment.

Information on chemicals associated with industrial processes may be obtained from standard reference sources such as Kirk-Othmer's *Encyclopedia of Chemical Technology* or the International Labor Office's *Encyclopedia of Occupational Safety and Health.*

3.1.3. Site Visits

The site visit is an essential element of the health assessment process. The site visit allows the assessor to observe firsthand the current conditions at the site. The assessor should note the current activities and land use at the site, public accessibility to the site, demographic characteristics of the community surrounding the site, and other information listed in Table 3.1. During the site visit, if possible, the assessor should meet with state or local health and environmental officials to discuss public health issues related to the site (see subsection 4.2.1.). The EPA project officer or representatives of the potentially responsible parties may provide information about ongoing or proposed remedial actions or other risk management strategies. Nearby residents and community action groups should be contacted for information on community health concerns related to the site. The site visit often provides the first opportunity for the community to meet with federal or state health professionals. Health assessors should describe the health assessment process and explain what the community can expect from a health assessment. The assessor should answer residents' questions and establish a foundation for a continuing dialogue with members of the community. During the site visit, the assessor may also visit other agencies to obtain private well inventories, geologic information, zoning restrictions, or other relevant information about the site.

3.2. COMMUNITY HEALTH CONCERNS

The assessor must attempt to identify specific community health concerns that are relevant to the site or release incident. Those concerns may be identified in the following ways:

1. **Community meetings** acquaint the assessor with the public and its health needs;

2. **Environmental and health complaints** made by the public and actions taken by local, state, and federal authorities may provide information on health concerns, enabling the health assessor to document past or current exposures; and

3. **Community health studies** may have been performed, indicating health effects as a result of exposure. Information about the type of study, relevance to the populations of concern at the site, study findings, and a possible contact person should be determined.

Additional guidance for gathering information on community health concerns can be found in Chapter 4. Populations need to be identified that may potentially become exposed to hazardous substances from the site; therefore, the assessor must have demographic information.

3.3. DEMOGRAPHIC INFORMATION

The size and characteristics of the populations most likely to have been exposed, and to be exposed, to contaminants must be determined. The health assessor should consider residential populations around the site, as well as individuals exposed at businesses, schools, and recreational areas near the site. Particular population information should be sought concerning the distance from the site to nearby residents and the size of the population within a specific radius of the site. Information on age distribution, ethnicity, and socioeconomic status of the affected community may assist in identifying susceptible subpopulations and in interpreting relevant health outcome parameters.

The primary source of U.S. demographic information is the Bureau of the Census of the Department of Commerce, which conducts and publishes a population census once every 10 years. The Bureau also provides other publications, tapes, and maps containing demographic data. Four particularly useful publications are: Number of Inhabitants (Series PC80-A), General Population Characteristics (Series PC80-1-B), Census Tracts (Series PHC80-2), and Block Statistics (Series PHC80-1). The Bureau's computer tapes, the Master Area Reference Files, provide useful census data.

Another useful source of census data is the Summary Tape File-1 (STF-1), which presents population data down to the block level. Data on those tapes are taken from the full census count rather than a sample of the population, making it the most accurate information available.

Printed publications of the Bureau of the Census are available from the Superintendent of Documents, U.S. Government Printing Office, Washington, DC 20402 (202) 783-3238. Publications on microfiche or tape can be ordered from Customer Services, Data User Services Division, Bureau of the Census,

Department of Commerce, Washington, DC 20233 (301) 763-4100.

Local government sources and health agencies can provide demographic information specific to the area to fully characterize the populations at risk. In addition, ATSDR is evaluating commercial sources of demographic information that may be used in the future.

Information about the residential, recreational, and occupational activities of potentially exposed populations should be identified. The types and levels of activities engaged in by populations at risk can affect frequency and duration of exposure.

3.4. LAND USE AND NATURAL RESOURCE USE INFORMATION

A review of land use at or near the site is necessary because it provides valuable information on the types and frequency of activities of the surrounding population and the probability for human exposure. Land use will significantly affect the types and frequency of human activities, thereby affecting the degree and intensity of contact with soils, water, air, exposed wastes, and consumable plants and animals. Past, present, and future land use should be considered.

Site accessibility and accessibility of the contaminated media are likely to affect the number of potentially exposed individuals. The presence, integrity, and suitability of fences and gates at the site should be determined, as well as the location of residential areas and signs of access. Worker access to contaminated areas should be determined.

Industrial areas near the site that may potentially contribute to the exposure-related body burden of workers and residents should be noted.

Residential areas and indicators of lifestyle factors that may influence exposures or health implications should be identified (low-income or poverty-level housing, scavenged drums and materials, gardens, livestock, and private wells).

Recreational areas such as parks, playgrounds, and beaches may increase exposures and provide additional routes of exposure.

Food-production areas and the market for those foods (home, local, regional, commercial, or subsistence) may influence exposure considerations.

Surface-water use (recreational, agricultural, and drinking-water supplies) will affect exposure considerations. Creeks, streams, and drainage ditches near residences may transport contaminants, resulting in increased exposure for children.

Groundwater use (municipal water supplies and private wells) may determine exposed populations. Information on the use of private wells may be obtained from local water utilities. When adequate information on private well use is not available, the assessor may suggest that a survey be conducted by the U.S. Geological Survey (USGS) or by the state or local health agency.

Hunting and fishing may affect exposure considerations for some populations. Information about hunting and fishing in the area can be obtained from the state fish and game department or local sporting goods stores.

After identifying land use and natural resource use information, the assessor should review information about contamination problems at the site.

3.5. ENVIRONMENTAL CONTAMINATION INFORMATION

Chemicals identified in specific environmental media represent the minimum environmental knowledge required to perform a preliminary health assessment. Therefore, additional information is needed to perform a more complete health assessment.

Chemical concentrations in each environmental medium are needed to determine the magnitude of exposure. In addition, sampling dates, locations, and methods are needed to determine representativeness, adequacy, and suitability of sampling information for a health assessment.

Quality assurance and quality control (QA/QC) information is necessary to identify adequacy of data quality for field and laboratory investigations. Detection limits and QA/QC information should be provided in a Data Review Summary. More extensive information on QA/QC procedures is presented in Appendix C.

The Toxic Chemical Release Inventory (TRI) contains information on the annual estimated releases of toxic chemicals into the environment. This information may be useful in the qualitative assessment of contamination found in on- or off-site environmental media. It should be noted that the TRI includes only chemical releases that have been reported since the database was initiated in 1987. The TRI is based on data collected by the EPA and is publicly accessible on the National Library of Medicine's (NLM) Toxicology Data Network (TOXNET).

Mandated by Title III of the Superfund Amendments and Reauthorization Act (SARA) of 1986, the TRI contains provisions for the reporting, by industry, on the releases of more than 300 toxic chemicals into the air, water, and land. Data submitted to EPA include names and addresses of facilities that manufacture, process, or otherwise use those chemicals, as well as amounts released into the environment or transferred to waste sites. Title III, also known as the Emergency Planning and Community Right-to-Know Act, calls for the EPA to collect this data nationwide, on an annual basis. The law mandates that the data be made available to the public through a computer database.

TRI data is arranged in the following broad categories:

- facility identification,

- substance identification,

- environmental release of chemical,

- waste treatment, and

- off-site waste transfer.

The data include the names, addresses, and public contacts of plants manufacturing, processing, or using the reported chemicals, the maximum amount stored on site; the estimated quantity emitted into the air (point and non-point emissions), discharged into bodies of water, injected underground, or released onto land; methods used in waste treatment and their efficiency; and data on the transfer of chemicals off site for treatment and disposal, either to publicly owned treatment works or elsewhere.

Health assessors should check the TRI for information on the release of hazardous substances from a site. If there are other facilities in the vicinity of a site, they should also be checked as possible sources of chemical releases.

In addition to these data needs, the assessor should identify information about environmental transport pathways.

3.6. ENVIRONMENTAL PATHWAYS INFORMATION

Physical characteristics and climatic conditions of the site contribute to the transport of contaminants and, ultimately, to human exposure to the contaminants. Therefore, information about physical and climatic characteristics of the site is needed.

Topography, the relative steepness of slopes and elevation of the site, may affect the direction and rate of water runoff, rate of soil erosion, and potential for flooding.

Soil types and locations influence percolation, groundwater recharge, contaminant release, and transport rates.

Ground cover of the site greatly influences the rates of rainwater infiltration and evaporation and soil erosion.

Annual precipitation affects the amount of moisture that is contained in the soil and the amount of percolation, as well as the water runoff and groundwater recharge rates.

Temperature conditions affect rates of contaminant volatilization and the frequency of human activity out-of-doors.

Other factors, such as wind speed, may influence volatilization and soil erosion.

Hydrogeological composition and structure affect the direction and extent of contaminant transport in water.

Locations of surface-water bodies and use of those water bodies may significantly affect the migration of contaminants off the site and into other media.

3.7. HEALTH OUTCOME DATA

An integral part of the evaluation of a site is the identification of relevant, site-specific health outcome data. Key health outcome data sources include state health departments and local public health officials. In addition, several health outcome databases are maintained by federal and private agencies. ATSDR is in the process of developing strategies to establish a baseline inventory of all existing relevant health outcome databases. Once developed, this consolidated database will be available to all health assessors.

Health outcome data constitute a key source of information for conducting health assessments. All health outcome databases and information used in the health assessment should be listed in this subsection. Chapter 7.2. provides guidance on the use of health outcome data in the health assessment process.

After reviewing the information necessary to characterize the site, the assessor may begin the next step in the health assessment process: identifying community health concerns (Chapter 4).

4. RESPONDING TO COMMUNITY HEALTH CONCERNS

Community health concerns constitute one of the three main pieces of information in health assessments. This chapter discusses the methods used to gather community concerns and how to respond to them during the health assessment process.

The health assessor works with the regional representatives and the Division of Health Assessment and Consultation's community involvement liaison, or other members of the site team, to gather and address community health concerns. Regional representatives play an important role in assisting the health assessor contact other federal agencies and state and local governments. The community involvement liaison facilitates contact between health assessors and the community; is a source of community-based information; serves as a source of information for the community about ATSDR and its activities; is responsible for announcing site-related meetings and the public comment period for health assessments; and may participate in site visits.

Before the on-site effort begins, the health assessor, community involvement liaison, regional representative(s), and other site team members should coordinate tasks and roles and determine their individual responsibilities. To identify and address community health concerns, the health assessment team undertakes the following tasks:

- identifying involved community members;

- involving the community in the health assessment process at the earliest opportunity;

- maintaining communication with the community and other involved parties throughout the process; and

- soliciting and responding to community comments on the final health assessment.

4.1. BEFORE THE SITE VISIT

4.1.1. Contacting Relevant Agencies

Sources of information about the site exist at the federal, state, county, and local levels. Depending on site-specific issues, the health assessor may need to communicate with health agencies and elected or appointed officials at each of those levels. Those organizations can facilitate community participation in the information-gathering process and play an important role in disseminating information to the community.

The health assessor should begin any information search by enlisting the aid of the regional representative or other site team members for assistance in contacting relevant agencies and discussing the types of assistance and information needed (e.g., access to site files, participation in site visits, review of draft health assessment documents, notices of public meetings being held, agency information on community networks, mailing lists).

4.1.1.1. Environmental Protection Agency

Well in advance of the site visit, through the regional representative or other members of the site team, the health assessor should contact the EPA for assistance in accomplishing the following tasks:

- identifying community contacts and existing information distribution channels;

- minimizing conflicting information;

- developing a plan to effectively use joint public meetings and other mechanisms to communicate with involved parties; and

- responding to community requests for information.

The health assessor, regional representative, or community involvement liaison should also contact the EPA remedial project manager and/or community relations staff to obtain the site-specific **community relations plan**. This plan, which the EPA prepares for every NPL site, includes site background data, a history of community involvement, community relations strategies and schedule, and a list of community contacts.

Two types of EPA site files—the **administrative record** and the **information repository**—will provide the health assessor with vital information. The administrative record contains documents that the EPA considered or relied on when selecting a response action. The information repository contains each document made available to or received from the public.

4.1.1.2. Other Agencies

Following contact with EPA, the health assessor may find additional site information at other organizations, including these:

- other federal agencies (e.g., U.S. Geological Survey, U.S. Forest Service, Indian Health Service, U.S. Fish and Wildlife Service, U.S. Department of Agriculture);

- state and local agencies (e.g., state public health and environmental departments);

- county medical office (including county medical officer, sanitarians, nurses, industrial hygienists); and the

- county environmental health department.

Health assessors should consult with the regional representative and, as needed, the ATSDR Washington, D.C., office to determine who should make these contacts.

4.1.2. Community Contacts

The community associated with a site can be broadly defined as the population living around the site and all others who can provide or disseminate relevant information on that site during the health assessment process. The involved community may include individual residents living near the site or organized community groups and their representatives. Contacting the population living around the site and interacting with community-based organizations allows the health assessor to become aware of community health concerns and to obtain other relevant site information.

Community concerns associated with exposure to site contaminants may be environmental or health-related, or both. The health assessment process focuses on obtaining information on health-related concerns. Identifying those concerns is critical and often requires active investigation by the health assessor. The first step is to identify as many relevant community contacts as possible. Determining who the appropriate community contacts are depends not only on site-specific issues, but also on the nature of the concerns and the degree to which the community is involved. A review of the site file may assist the health assessor in identifying other persons—both within and outside the agency—who have been involved with the site.

Key community representatives identified for contact should receive information from the health assessment team about the upcoming site visit and requests for meetings with community members. Those community contacts should be able to provide the health assessor and other members of the site team with valuable information about the best ways to secure site data and information about community concerns, the level of community interest, and the best strategy for interacting with the site community (e.g., public meetings, public availability meetings, small group meetings). The health assessor can begin determining the extent of concern within the community by noting the nature and number of questions residents ask.

As appropriate, the health assessor should contact individuals and established community groups, including (but not limited to) these:

- Individual residents/petitioner(s);

- Elected officials (U.S. Congress, state, and local officials). Appropriate ATSDR channels must be used and applicable government regulations must be followed when contacting elected officials;

- Fishing, hunting, agricultural, conservation, and industrial organizations;

- Local medical society and other health care providers;

- Media (print, electronic);

- Community leaders;

- Community organizations, including specialized minority organizations;

- Local community environmental groups;

- Universities or academic institutions;

- School principals and school nurses;

- Labor unions;

- Institutions and facilities near the site (e.g., child-care centers, prisons); and

- Potentially responsible parties .

4.1.3. Site Visit Communication Strategy

Based on the information collected, the health assessor works with others involved with the site to develop a specific strategy for communicating with the community during the site visit. In cooperation with the regional representative, community involvement liaison, or other staff, the health assessor should establish a meeting schedule and determine the type of meeting most suited to the needs of the community. A public meeting, an availability meeting, or a small-group meeting (but generally not all three) may be held; individual

meetings with key contacts should always occur.

4.1.3.1. Media Relations and Community Notification

At some sites, ATSDR may announce the site visit to the local media with a press release. The press release discusses the purpose and uses of a health assessment, the purpose of the site visit, and any meetings to which the public is invited.

If a public availability session or public meeting is to be held, ATSDR issues a press release and places a public notice in the local newspaper; fliers may be sent to key contacts requesting that they be distributed in the community. As needed, notices should also be sent to local commercial establishments requesting that they be posted.

4.2. DURING THE SITE VISIT

The site visit is a key step in the health assessment process. The primary purposes of the site visit are to observe the site and meet with key agencies to gather information on the community's health concerns. Site observations should include information based on interviews with local, state, and federal environmental and public health officials and community members who have knowledge of the site.

A successful site visit requires thorough preparation and organization. Before the site visit, the health assessor, regional representative, the community involvement liaison, and other members of the site team should meet and make arrangements to:

- brief all contacts about the purpose of the visit;

- send the contacts written confirmation of site visit and meeting dates, times, and places;

- make arrangements for group and public meetings or ensure those duties have been assigned;

- invite representatives of relevant agencies (EPA, state health and environmental departments) to appropriate meetings or visits; and

- develop informational materials (press releases, fact sheets).

Activities conducted during the site visit will vary from site to site; not every site visit will include a public meeting or a live media interview. Chapter 3 addresses in detail the role of the site visit in the health assessment process.

4.2.1. Meetings

Initial contacts made during the site visit set the tone for the agency's continued involvement with the community. For that reason, smaller meetings are preferred over large-scale public meetings, which do not offer one-on-one contact. Initial meetings with the community serve a variety of purposes, including these:

- further identifying the concerned community (including demographic and geographic distribution information) and the community's concerns (including specific health outcomes of concern and quality-of-life issues);

- gathering information on past and present community interest in the site;

- further identifying key contact people within the concerned community;

- learning about site status and the community's perception of it;

- determining possible exposure routes and the potential for exposure;

- building community trust;

- educating the community about ATSDR and its activities (identifying what the agency can and cannot do in the context of a health assessment and stressing the

non-regulatory nature of the document and its advisory capacity) and the purpose, scope, and possible results of a health assessment;

- providing assistance to community members who want to better understand technical issues regarding site-related contaminants and exposure and providing information about other sources of information (such as ATSDR Toxicological Profiles, national environmental groups, nearby university centers, etc.);

- developing a mailing list;

- identifying candidates for a Community Assistance Panel (panels serve as a conduit of information between ATSDR and the community, providing the community the opportunity to become involved in the health assessment process);

- identifying ways the community prefers to receive information (e.g., through the news media, a quarterly newsletter, etc.) and any established communication frameworks (e.g., schools, churches, city government, and/or informal means); and

- identifying how the community wishes to be involved in the health assessment process.

4.2.2. Media Contact

Reporters may attend public meetings, which are large-scale, open forum events. Public availability sessions, on the other hand, are not media opportunities. These meetings are intended to be informal, private sessions at which individuals from the community can share their concerns and ask questions. Many states have a *sunshine law*, which prohibits barring reporters from these meetings. Should reporters arrive, staff should meet with them; stress that the session is to allow citizens privacy in discussing their concerns, and request a later appointment to answer media questions. To protect the community's privacy, filming and/or interviewing participants should be discouraged.

To ensure that the emphasis of any public meeting remains on the community and its concerns, the health assessor may wish to announce at the beginning of the meeting that the session is for the community, and media questions will be taken afterward.

4.3. AFTER THE SITE VISIT

4.3.1. Documenting and Sharing Information

Information obtained during the site visit must be documented by a site visit report and an ATSDR Record of Activity.

Persons involved in the site visit should meet as soon as possible following the visit to discuss community involvement activities. Specifically, the need for a Community Assistance Panel should be determined.

Information gathered before, during, and after the site visit is used to prepare a draft Community Involvement Plan (see 4.3.2.), which may be revised as necessary.

4.3.2. The Site-Specific Community Involvement Plan

ATSDR prepares a site-specific community involvement plan for each site that describes ATSDR's goals, objectives, and strategies for involving the community in the health assessment. In addition, this plan may include the following elements:

- information repository;*
- quarterly updates;*
- Community Assistance Panel; and/or
- follow-up community meetings.

*These activities will be piloted during 1991-1992 and evaluated for effectiveness.

4.4. PUBLIC COMMENT PERIOD

Releasing a health assessment for public comment is the last step in this phase of the process. The purposes of the public comment period are:

- to provide the public, particularly the community associated with the site, the opportunity to comment on the health assessment, especially the public health conclusions and recommendations and the effectiveness of the document in addressing community health concerns; and

- to provide ATSDR with additional relevant information.

The health assessment is available in the community at the established information repository (if one exists) and at two other repositories, if possible. To facilitate distribution of the draft health assessment and to encourage public participation, certain groups (e.g., local health department and community contacts) should receive copies. The document's availability is announced in a legal public notice placed in one or more local newspapers and in a press release to local media. In addition, letters and/or fliers about the document may be sent to key community contacts, and the community may also be notified by other methods previously identified as effective (such as notices sent home through the schools or civic organizations, telephone *trees*, etc.)

A log of comments received is kept as part of the health assessment administrative record. Comments (without attribution) and agency responses are included as an appendix to the completed health assessment. For each site, a separate file will be created containing the health assessment, the public comments, and corresponding responses. The file is available for public inspection upon written request.

The health assessor must respond to public comments, revising the health assessment when appropriate. Following the close of the public comment period, the health assessor(s) and other parties involved in the health assessment meet to discuss the need for a public or other community meeting to announce the results of the health assessment. Criteria for assessing the appropriateness of a public meeting include these:

- number of comments received (as an estimate of community interest);

- advice of members of the Community Assistance Panel, if applicable;

- input from key community contacts and/or the expressed wishes of the larger community as indicated in meetings, by telephone contacts, or other means;

- amount and type of media coverage;

- history of community interest--estimated by a variety of factors, such as number of community-based environmental groups, numbers of persons visiting the information repository and/or providing information through that mechanism, or calls to ATSDR staff from community members; and

- number of persons who have attended meetings.

If a public meeting is to be held, the health assessor should be prepared to discuss responses to public comments as well as other issues associated with the health assessment.

4.5. ADDING COMMUNITY HEALTH CONCERNS TO THE HEALTH ASSESSMENT

Community health concerns should be discussed in two parts of the health assessment: the Community Health Concerns and Public Health Implications sections. In the Community Health Concerns section, the health assessor presents the health concerns of the community residents, but no assessment or evaluation of the concerns is provided in that section of the health assessment. ATSDR's response to health concerns should be included in the Public Health Implications section under the subsection Community Health Concerns Evaluation. Sections 9.3. and 9.6.3. of this manual describe in detail how health assessors should present information on community health concerns.

5. DETERMINING CONTAMINANTS OF CONCERN

This chapter discusses how to determine contaminants of concern at the site. Contaminants of concern are the site-specific chemical substances that the health assessor has selected for further evaluation of potential health effects. Identifying contaminants of concern is an iterative process that requires the assessor to examine contaminant concentrations at the site, the quality of environmental-sampling data, and the potential for human exposure. The following kinds of information will assist the assessor in identifying contaminants of concern:

1. **Contaminants on and off the site** (Although all contaminants should be considered potential contaminants of concern, health assessors select contaminants of concern based on comparative analyses with health guidelines, multi-media exposures, interactive effects, and community health concerns.)

2. **Concentrations of contaminants in environmental media** (Health assessors use sampled data—temporal analyses, and spatial analyses, when possible—to identify contaminants of concern in the past, present, and future, and the likelihood of inter-media transfer.)

3. **Background concentration levels** (A review of background contaminant levels in the local environmental media should assist in identifying the source of contamination.)

4. **Quality of environmental-sampling data and techniques** (Sampling data and techniques should be evaluated for validity and representativeness.)

5. **ATSDR Environmental Media Evaluation Guides (EMEGs) and other appropriate comparison values** (Those environmental concentration guides are used to help select contaminants of concern.)

6. **Community Health Concerns** (A health assessor must address each community health concern about a particular contaminant, regardless of its presence or concentration at the site.)

7. **Toxic Chemical Release Inventory** (This EPA database must be examined to determine additional sampling needs and additional sources of contamination in the area, as well as the dates, amounts, and names of contaminants that have been released by the site facility and others in the vicinity.)

8. **ATSDR Toxicological Profiles** (These documents provide a public health statement, health effects information, chemical and physical properties, use information, exposure information, analytical methods, regulations, and references.)

5.1. IDENTIFYING CONTAMINANTS

The reports and documents made available to the assessor generally mention the contaminants found in environmental media on and off the site. Those reports usually provide appendices that list the media involved, sample number, date sampled, detection limits, and concentrations detected. The list of detected contaminants may be quite extensive, involving many compounds. The health assessor should consider all detected contaminants as potential contaminants of concern.

5.1.1. Organization of Contaminants of Concern

The health assessor should identify, organize, and discuss the contaminants of concern by media, keeping the narrative discussion to a minimum. While the narrative should summarize information in the data tables, the concentrations should not be repeated. Media subheadings, such as Surface Soil, Subsurface Soil, Sediment, Surface Water, Groundwater (with subcategories of Private Wells, Public Wells and Monitoring Wells), Air, Biota, Waste Materials, Soil Gas, and Leachate should be used when appropriate in the On-site and Off-site Contamination subsections.

All contaminants identified in the On-Site Contamination subsection should be included in the Off-site Contamination subsection. The health assessor should identify on-site contaminants that have not been detected or reported off site. Regardless of whether the data were available for review, all environmental media that have been sampled on- or off-site should be explicitly stated in the health assessment.

Data on surface and subsurface soil should be separated, as should groundwater data from private wells, public wells, and monitoring wells.

When discussing surface water data, the health assessor should clearly differentiate between puddles and surface waters found in larger water bodies and waterways. Surface waters do not include impoundments or lagoons that contain waste materials.

The health assessor should address trends in discussions of the data, but should not discuss migration in this section. Specifically, the health assessor should consider spatial distribution, "hot spots," concentration changes over time, and contamination differences between media.

5.1.2. Presenting Contaminants of Concern in Data Tables

When a substance is identified as a contaminant of concern in one medium, the health assessor should provide its concentration for all media sampled in either a data table or the narrative. Unless the data table is too long, the health assessor should incorporate the data table into the narrative portion for each environmental medium. This should be done for both the On-site Contamination and the Off-site Contamination subsections. Comparison values should be included in the data tables. For specific examples on how to construct data tables and on which items to include, refer to Chapter 9, Health Assessment Format.

The range of contaminant concentrations detected should be presented in the data tables. Averaged data can also be reported, if available, but the average should be accompanied by the range of concentrations. For the purpose of selecting contaminants of concern, the maximum concentration of a contaminant should be used. This ensures that all potentially significant contaminants will be evaluated.

To distinguish historical data from current data, the health assessor should identify in a data table contaminant concentrations that document past exposures. It will be necessary for the health assessor to determine whether past removal or remedial activities may have altered the contamination in a medium. The health assessor should not rule out past or present contamination or transport until the following occurs:

1. It is determined that the sampling design, coverage, location or station construction, collection and frequency, sample storage and shipment, or analyses are satisfactory.

2. Information is found explicitly stating that removal or remedial activities have occurred in the sampled medium.

5.2. REVIEWING CONCENTRATION LEVELS

The health assessor needs to review the concentration levels reported for each contaminant. In addition to listing contaminants detected, site reports and documents contain data on contaminant concentrations at the source, in areas of special concern, and, in some cases, at background locations. Review of this information can provide the assessor with preliminary indications of those contaminants detected at the highest concentration levels and with the greatest frequency. Further, the assessor may check sampling completeness and representativeness by identifying the specific media that were sampled and the exact location of sampling points.

Background locations represent areas at or near the hazardous waste site that are not contaminated. Refer to Section 5.5. for more details on obtaining and using background data. At these background locations, samples may be collected from each medium that has the same basic characteristics as the medium of concern at the site. Determining background locations requires knowledge about the site history, site spills and releases, and about which directions relative to the site, or its contamination plumes, are upslope, upgradient, upwind, or upstream.

5.3. EVALUATING SAMPLING DATA AND TECHNIQUES

Before using environmental data to reach conclusions in the health assessment, the assessor must check for inadequacies, insufficiencies, and discrepancies in the data or in the sampling and analytical techniques used to obtain the data. That is done by obtaining the QA/QC summary from EPA through the ATSDR regional representative. Once obtained, the health assessor should verify the acceptability of three data criteria: field data quality, laboratory data quality, and data adequacy.

5.3.1. Field Data Quality

The health assessor should verify that the sampling data contained in the site evaluation report were obtained in accordance with QA/QC specifications, as detailed in the site's Quality Assurance Project Plan, to ensure appropriate sampling techniques and protocol. The QA/QC summary in the EPA data package should state if sampling quality is not adequate. When available, the assessor should review information about these elements:

- field inspections;

- well development and installation;

- bottles and preservatives;

- decontamination procedures;

- equipment calibration; and

- field duplicates and splits.

5.3.2. Laboratory Data Quality

These criteria may be verified by reviewing the case narrative and the Data Review Summary that should be provided with the QA/QC summary in the EPA data package. Similar information should be available for both Contract Laboratory Program (CLP) and non-CLP data.

The **case narrative** is prepared by the laboratory performing the data analysis. This narrative contains a summary of any QC, sample, shipment, or analytical problems, and documentation of the internal decision process used. It outlines problems encountered and their final solutions.

The **Data Review Summary** is prepared by EPA regional staff. This summary documents the validation of sample holding times, instrument performance, calibration, blanks, surrogate recovery, matrix spike recovery, and compound identification. It includes documentation of actions taken to resolve data quality problems and an overall case assessment.

If QA/QC information is not available, the assessor should state that the conclusions drawn for this health assessment are determined by the availability and reliability of the referenced information, and it is assumed that adequate quality assurance and quality control measures were followed with regard to chain of custody, laboratory procedures, and data reporting. When QA/QC information is not available, the assessor should account for the uncertainty of the data in both the conclusions and recommendations of the health assessment.

5.3.3. Adequacy of Data

After reviewing the media-specific concentration data and field and laboratory data quality information, the health assessor must determine whether the information available for each medium is sufficient to make an assessment of the levels of contaminants to which people are or might be exposed. If the data are insufficient or not available, then a brief explicit statement should be made in the Recommendations section of the health assessment report specifying the information needed.

At NPL sites, environmental media are usually analyzed for chemical substances on the Target Compound List (TCL). That list includes most of the organic and inorganic chemical contaminants that are found at hazardous waste sites. However, no list or series of analytical tests can include all potentially hazardous chemicals. At some sites, operations conducted at the facility may have caused environmental contamination with chemical substances that are not on the TCL. If such contamination is suspected and is a source of potential health concern, the assessor should note the deficiency in this section and in the Conclusions section and make the appropriate recommendations in the Recommendations section to fill the data gap.

Inadequacies, insufficiencies, and discrepancies in the data or in the sampling and analytical techniques used to obtain the data, as well as missing QA/QC information, should be explicitly noted in the health assessment. Additional information on QA/QC procedures is presented in Appendix C.

Before making comparisons with site-related or background data, the data's representativeness and adequacy should be evaluated. The health assessor should consider the following items:

1. Are the data representative of the media of interest for a particular area and time period (e.g., point of time, season, year, or decade)?

2. Do sufficient data exist to understand spatial or temporal trends and the potential points of exposure?

The health assessor should address the following questions on type and design of environmental sampling to determine the adequacy of the data.

Soil

If humans have access to contaminated soils, can ranges of contamination be provided on the basis of land use (i.e., restricted access, road/driveway/parking lot access, garden use, agriculture and feedlot use, residential use, playground and park use, etc.)?

Have the soil depths been specified? Do soil data represent "surface soil" data (≤ 3 inches in depth) or "subsurface soil" data (>3 inches in depth)? If soil depth is known, but does not meet surface or subsurface soil definitions, designate the data as *soil* and specify the depth (e.g., 0-6 inches). If the soil depth is unknown, the health assessor should designate the data as unspecified soil."

Has soil been defined in the data? If not, the health assessor should assume soil includes any unconsolidated natural material or fill above bedrock that is not considered to be soil and excludes manmade materials such as slabs, pavements or driveways of asphalt, concrete, brick, rock, ash or gravel. A soil matrix may consist of pieces of each of these materials.

Do soil data include uphill and downhill samples and upwind and downwind samples both on and off the site?

Sediment

Have the sediment samples been identified as grab samples or cores? Was the depth of the samples specified?

Was the sampling program designed to collect sediment samples at regular intervals along a waterway or from depositional areas or both?

Do the sediment data include results for upstream and downstream samples both on- and off-site?

Has sediment been defined by the samplers? (To prevent confusion between sediment and soil, assume "sediment" is defined as any solid material, other than waste material or waste sludge, that lies below a water surface, that has been naturally deposited in a waterway, water body, channel, ditch, wetland, or swale, or that lies on a bank, beach, or floodway land where solids are deposited.)

Have any sediment removal activities (e.g. dredging, excavation, etc.) occurred that may have altered the degree of sediment contamination (leading to a false negative). This becomes important when the following occur:

1. sediment contamination in fishable waters is used to justify sampling and analyses of edible biota;

2. sediment data are used to justify additional downstream sampling, particularly at points of exposure and in areas not subject to past removal activities; and

3. the significance of past exposure is assessed.

Surface Water

Do surface-water data include results for samples both upstream and downstream of the site?

Was information obtained on the number of surface-water samples taken at each station, as well as the frequency, duration, and dates of sampling?

Groundwater

Were groundwater samples collected in the aquifer of concern?

Did sampling occur both upgradient and downgradient of the site and the site's groundwater contamination plume?

Other Comments

Did the sampling design include selected hot spot locations and points of possible exposure?

5.4. MAKING DATA COMPARISONS

Health assessors may perform either a direct comparison or a statistical comparison between one data set and another, or both. The simple, direct comparison between one datum (e.g., a relevant background datum) and one or more site-related datum is the most popular approach.

The health assessor may also perform a statistical comparison to examine the significance of any differences between various site-related data (e.g., well data at contamination source versus downgradient well data at the site boundary) or between site-related data and background data. For an overview of basic statistical concepts in comparing background data with site-related contamination, refer to background sampling in EPA's *Risk Assessment Guidance for Superfund Volume I - Human Health Evaluation Manual*

(Part A), December 1989. This EPA document explains how to perform statistical hypothesis testing to determine whether a significant difference exists between contaminant concentrations at the site and background data.

Often, statistical analyses are unnecessary because site-related data and background data clearly differ. A more important issue will be determining the representativeness of the sampling data at the site and at background locations. See Section 5.3., Evaluating Sampling Data and Techniques.

5.5. USING BACKGROUND CONCENTRATIONS

Although comparisons of site-related data with background data may influence the selection of a contaminant of concern, background levels must not be considered the sole determinant. After background sampling data have been found to be representative, the health assessor should determine whether the background data exceed Environmental Media Evaluation Guides (EMEGs; section 5.6.) and other applicable comparison values. That comparison of background data with EMEGs and other comparison values will assist the health assessor in determining whether any health hazards or threats may be posed by background levels. Background levels in local environmental media may be greatly influenced by high levels of contaminants from native mineral deposits or other natural sources.

Combined with other site-related sampling data, local background data may be used by the health assessor to help determine the likely source of contamination. Both spatial and temporal analyses of the data points can assist in this determination.

5.5.1. Kinds of Background Data

Background data may be one or a combination of two types of levels:

1. **naturally occurring** ambient levels of substances in the environment that have not been influenced by humans; and

2. **anthropogenic** levels of substances in the environment due to human-made, non-site sources.

Background levels can be localized or ubiquitous depending upon the extent and source of contamination. Background levels may be obtained locally from each medium of concern, in areas not influenced by site contamination. For example, as background data on indoor air and dust levels and building materials become available, health assessors can compare that information to indoor air and dust levels and building materials of structures (e.g., residences) contaminated by site contaminants.

If local background samples for the medium of concern have not been collected, background levels for the state, region, or nation might be used for comparison provided the medium has the same basic characteristics as the medium of concern at the site. If sources other than local data are used for background data, reference the source and explain their applicability.

Background levels of many chemicals in various environmental media are discussed in ATSDR's Toxicological Profiles. In addition, Table 5.1. contains ranges of background levels for elements in soils of the United States. The health assessor may use that table for comparison purposes if local background soil data are not available.

5.5.2. Background Comparisons With EMEGs or Other Comparison Values

Whenever local background data exceed EMEGs or applicable comparison values, the health assessor should:

1. further evaluate the contaminant in subsequent sections of the health assessment; and

Table 5.1. Mean Concentrations[1], Deviations, and Ranges of Elements in Samples of Soils in the Coterminous United States[2]

	Western United States (west of 96th meridian)				Eastern United States (east of 96th meridian)			
Element	Mean	Deviation	Observed Range	Estimated Arithmetic Mean	Mean	Deviation	Observed Range	Estimated Arithmetic Mean
Al%	5.8	2.00	0.5 - >10	7.4	3.3	2.87	0.7 - >10	5.7
As	5.5	1.98	<0.10 - 97	7.0	4.8	2.56	<0.1 - 73	7.4
B	23	1.99	<20 - 300	29	31	1.88	<20 - 150	38
Ba	580	1.72	70 - 5,000	670	290	2.35	10 - 1,500	420
Be	0.68	2.30	<1 - 15	0.97	0.55	2.53	<1 - 7	0.85
Br	0.52	2.74	<0.5 - 11	0.86	0.62	2.18	<0.5 - 5.3	0.85
C%	1.7	2.37	0.16 - 10	2.5	1.5	2.88	0.06 - 37	2.6
Ca%	1.8	3.05	0.06 - 32	3.3	0.34	3.08	0.01 - 28	0.63
Ce	65	1.71	<150 - 300	75	63	1.85	<150 - 300	76
Co	7.1	1.97	<3 - 50	9.0	5.9	2.57	<3 - 70	9.2
Cr	41	2.19	3 - 2,000	56	33	2.6	1 - 1,000	52
Cu	21	2.07	2 - 300	27	13	2.8	<1 - 700	22
F	280	2.52	<10 - 1,900	440	130	4.19	<10 - 3,700	360
Fe%	2.1	1.95	0.1 - >10	2.6	1.4	2.87	0.01 - >10	2.5
Ga	16	1.68	<5 - 70	19	9.3	2.38	<5 - 70	14
Ge	1.2	1.32	0.58 - 2.5	1.2	1.1	1.45	<0.1 - 2.0	1.2
Hg	0.046	2.33	<0.01 - 4.6	0.065	0.081	2.52	0.01 - 3.4	0.12
I	0.79	2.55	<0.5 - 9.6	1.2	0.68	2.81	<0.5 - 7.0	1.2
K% 3	1.8	0.71	0.19 - 6.3	none	1.2	0.75	0.005 - 3.7	none
La	30	1.89	<30 - 200	37	29	1.98	<30 - 200	37
Li	22	1.58	5 - 130	25	17	2.16	<5 - 140	22
Mg%	0.74	2.21	0.03 - >10	1.0	0.21	3.55	0.005 - 5	0.46
Mn	380	1.98	30 - 5,000	480	260	3.82	<2 - 7,000	640
Mo	0.85	2.17	<3 - 7	1.1	0.32	3.93	<3 - 15	0.79
Na%	0.97	1.95	0.05 - 10	1.2	0.25	4.55	<0.05 - 5	0.78
Nb	9.7	1.82	<10 - 100	10	10	1.65	<10 - 50	12
Nd	36	1.76	<70 - 300	43	46	1.58	<70 - 300	51
Ni	15	2.1	<5 - 700	19	11	2.64	<5 - 700	18
P	320	2.3	40 - 4,500	460	200	2.95	<20 - 6,800	360
Pb	17	1.8	<10 - 700	20	14	1.95	<10 - 300	17
Rb	69	1.5	<20 - 210	74	43	1.94	<20 - 160	53
S%	0.13	2.37	<0.08 - 4.8	0.19	0.10	1.34	<0.08 - 0.31	0.11
Sb	0.47	2.15	<1 - 2.6	0.62	0.52	2.38	<1 - 8.8	0.76
Sc	8.2	1.74	<5 - 50	9.6	6.5	1.90	<5 - 30	8.0
Se	0.23	2.43	<0.1 - 4.3	0.34	0.30	2.44	<0.1 - 3.9	0.45
Si% 3	30	5.7	15 - 44	none	34	6.64	1.7 - 45	none
Sn	0.90	2.11	<0.1 - 7.4	1.2	0.86	2.81	<0.1 - 10	1.5
Sr	200	2.16	10 - 3,000	270	53	3.61	<5 - 700	120
Ti%	0.22	1.78	0.05 - 2.0	0.26	0.28	2.00	0.007 - 1.5	0.35
Th	9.1	1.49	2.4 - 31	9.8	7.7	1.58	2.2 - 23	8.6
U	2.5	1.45	0.68 - 7.9	2.7	2.1	2.12	0.29 - 11	2.7
V	70	1.95	7 - 500	88	43	2.51	<7 - 300	66
Y	22	1.66	<10 - 150	25	20	1.97	<10 - 200	25
Yb	2.6	1.63	<1 - 20	3.0	2.6	2.06	<1 - 50	3.3
Zn	55	1.79	10 - 2,100	65	40	2.11	<5 - 2,900	52
Zr	160	1.77	<20 - 1,500	190	220	2.01	<20 - 2,000	290

1   Means and ranges are reported in parts per million (mg/g) or percent, as indicated. Means and deviations are geometric except as indicated.
2   Source: U.S. Geological Survey (1).
3   Means are arithmetic, deviations are standard.
%   Percent.

2. indicate that the background contaminant is unrelated to the site, provided sufficient evidence is available.

5.5.3. Background Comparisons With Site-related Concentrations

Regardless of whether local background data exceed or do not exceed EMEGs and other applicable comparison values, the health assessor should compare site-related concentrations to (1) background data and (2) EMEGs and other health assessment comparison values. The first part of this comparison and a spatial analysis will help the health assessor identify the likely sources of site contamination.

When background data are compared with site-related concentrations, the following decisions **should be considered:**

1. If the levels for a site-related contaminant in an environmental medium are greater than background levels and less than EMEGs or other relevant comparison values, the medium may need to be further assessed for possible contaminant migration (in the Pathways Analyses section), especially if the extent of contamination in that medium has not been adequately characterized. In most cases, the contaminant should not be listed as a contaminant of concern in that medium. However, the health assessor

Table 5.2. Summary Table Of Decisions To Consider When Using Background Data And Site-Related Data

IF ON-SITE OR OFF-SITE DATA LEVEL IS	GREATER THAN BACKGROUND DATA LEVEL	LESS THAN BACKGROUND DATA LEVEL	GREATER THAN EMEG OR OTHER COMPARISON VALUE	LESS THAN EMEG OR OTHER COMPARISON VALUE	THEN LIST AS CONTAMINANT OF CONCERN
1. "	YES			AND YES	NO
2. "	YES		AND YES		YES
3. "	YES		NO EMEG	NO EMEG	YES
4. "		YES		AND YES	NO
5. "		YES	AND YES		YES
6. "		YES	NO EMEG	NO EMEG	YES

Example of how to read #1 on this table: If on-site or off-site data level is "greater than background data level" (marked with YES) and less than EMEG or other comparison value (marked with YES), then do not "list as a contaminant of concern" (marked with NO).

should use professional judgment in this determination, particularly if the medium has been poorly characterized.

2. If the levels for a site-related contaminant in an environmental medium are greater than background levels and EMEGs or other relevant comparison values, the contaminant should be listed as a contaminant of concern in that medium.

3. If the levels for a site-related contaminant are greater than background levels, and no EMEG or other applicable comparison value exists, the contaminant should be listed as a contaminant of concern.

4. If the levels for a site-related contaminant are less than background levels and EMEGs or other applicable comparison values, the contaminant should not be listed as a contaminant of concern. However, other factors such as multi-media exposures, interactive effects, or community health concerns may require selecting the contaminant as a contaminant of concern.

5. If the levels for a site-related contaminant are less than background levels, but greater than EMEGs or other applicable comparison values, the contaminant should be listed as a contaminant of concern.

6. If the levels for a site-related contaminant are less than background levels, and no EMEGs or other applicable comparison values exist, the contaminant should be listed as a contaminant of concern.

5.6. COMPARING ENVIRONMENTAL CONCENTRATIONS TO EMEGS

Health assessors are often presented with voluminous quantities of environmental sampling data for a site. Those data typically consist of analytical chemical determinations of the concentrations of contaminants in groundwater, surface water, soil, sediment, air, and sometimes other media. The health assessor must evaluate those data to determine which contaminants in which media pose potential health hazards. Environmental Media Evaluation Guidelines (EMEGs) have been developed to assist health assessors in this selection process. Using the EMEG values provides health assessors with a consistent strategy for selecting environmental contaminants that need to be further evaluated for potential health effects.

EMEGs are media-specific comparison values used to select chemical contaminants of potential concern at hazardous waste sites. These EMEG values have been developed for water, soil, and air. EMEGs vary as a function of exposure to contaminated media. Therefore, EMEG values are presented as a range of values, rather than as a single value. This range spans the exposure potential for different segments of the population. The health assessor should select the EMEG value that corresponds to the most sensitive segment of the population that is potentially exposed to contamination from the site.

The derivation of EMEGs is described in Appendix A of this manual. ATSDR EMEGs are based on the Minimal Risk Levels (MRLs) presented in the ATSDR Toxicological Profiles. At this time, MRLs consider only the non-carcinogenic toxic effects of a chemical substance. However, some chemicals have carcinogenic, as well as noncarcinogenic, toxicity. Therefore, for chemicals that are classified as carcinogens, a comparison value based on carcinogenic toxicity should be developed. To derive this value, health assessors can use the EPA's Cancer Slope Factors described in Appendix B. A dose of the carcinogen that corresponds to a 10^{-6} increase in the lifetime risk of cancer should be used in place of the MRL to calculate the comparison value. The methodology for this calculation is presented in Appendix A, Section A.5.

When evaluating environmental sampling data, the concentration of a contaminant should be compared to the appropriate EMEG (or to a comparable value). If the concentration of the contaminant is in excess of the EMEG, potential exposures to that chemical should be further evaluated for their health effects. If the concentration of a chemical is below the EMEG, then it is unlikely that exposures to that chemical in that medium would pose a public health hazard.

However, other factors such as multiple exposures, synergistic effects, and community health concerns may require including that chemical as a contaminant of concern.

The EMEG values **should not** be used as a predictor of adverse health effects or for setting cleanup levels. Their purpose is to provide health assessors with a means of selecting environmental contaminants for further evaluation. The application of EMEGs is an early step in the health assessment process, which must also include an evaluation of site-specific exposure pathways, community health concerns, and health outcome data. Chapter 7 provides further discussion on how to assess the public health implications of exposures to environmental contamination.

A list of EMEG values will be maintained in the ATSDR Hazardous Substances Data Management Systems. That list will be continually updated to add EMEGs for new chemicals and to revise existing ones as warranted by new toxicologic information.

5.7. COMMUNITY HEALTH CONCERNS

A health assessor must address each community health concern about a particular contaminant, regardless of its presence or concentration on the site. The health assessor will not address the community health concerns about a contaminant in this section, but will list the contaminant as one of concern. It is not necessary for the selected contaminant to exceed ATSDR's EMEGs or other relevant comparison values. The health assessor will indicate what the community health concern is in the Community Health Concerns section and will address and discuss the concern in the Community Health Concerns Evaluation subsection of the Public Health Implications section.

5.8. TOXIC CHEMICAL RELEASE INVENTORY

Health assessors must use the Toxic Chemical Release Inventory (TRI), an EPA database, as a tool for determining the following information:

- additional sampling needs;

- additional sources of contamination in the area;

- amounts and names of contaminants that have been released by the site facility and others in the vicinity;

- analysis of contaminants not found on the EPA Target Compound List; and/or

- additional contacts for site information, environmental data, community health concerns, and health outcome data.

5.9. REFERENCES

1. Shacklette HT, Boerngen JG. Element concentrations in soils and other surficial materials of the conterminous United States. U.S. Geological Survey Professional Paper 1270. Washington, DC: U.S. Government Printing Office, 1984.

2. *Federal Register*, Vol. 51, No. 185, pages 34014-34025, Wednesday, September 24, 1986.

3. National Research Council. Complex mixtures. Washington, DC: National Academy Press, 1988.

4. U.S. Department of Health and Human Services. Fifth Annual Report on Carcinogens - 1989. Washington, DC: U.S. Government Printing Office, 1989.

6. IDENTIFYING AND EVALUATING EXPOSURE PATHWAYS

The purpose of this chapter is to show the health assessor how to (1) identify each of the five elements of an exposure pathway; (2) determine whether these elements are linked to form an exposure pathway; (3) categorize an exposure pathway as a completed exposure pathway or a potential exposure pathway; and (4) determine whether the exposure pathway should be eliminated or further discussed in the health assessment. This chapter also discusses the preferred and optional approaches for presenting and discussing exposure pathways in the Pathways Analyses section of the health assessment.

An exposure pathway is the process by which an individual is exposed to contaminants that originate from some source of contamination. An exposure pathway consists of the following five elements:

1. **Source of contamination** (source of contaminant release into the environment, **or** the environmental media responsible for causing contamination at a point of exposure if the original source of contamination is unknown);

2. **Environmental media and transport mechanisms** (environmental media include waste materials, groundwater, surface water, air, surface soil, subsurface soil, sediment, and biota. Transport mechanisms serve to move contaminants from the source to points where human exposure can occur);

3. **Point of exposure** (a location of potential or actual human contact with a contaminated medium, e.g., residence, business, residential yard, playground, campground, waterway or water body, contaminated spring or hand-drawn well, food services, etc.);

4. **Route of exposure** (means by which the contaminant actually enters or contacts the body, such as ingestion, inhalation, dermal contact, and dermal absorption); and

5. **Receptor population** (persons who are exposed or potentially exposed to the contaminants of concern at a point of exposure).

Those five elements are illustrated in Figure 6.0.

The health assessor should include in the health assessment only information that is necessary for the development and understanding of the exposure pathways. The health assessor should exclude any information not essential to the exposure pathway being assessed. Conclusions about any health threats should not be made in the Pathways Analyses section.

The assessor should note that an exposure pathway is not simply an environmental medium (e.g., air, soil, groundwater, or surface water) or a route of exposure. Rather, an exposure pathway includes all the elements that link a contaminant source to a receptor population. The elements of an exposure pathway may occur in the past, present, or future.

Different routes of exposure to a contaminant may result in different health concerns. A specific medium or exposure route may be part of multiple exposure pathways, and different transport mechanisms may result in persons being exposed to different contaminant levels. For example, consider a scenario in which a shallow water table aquifer transports contaminants to private residential wells near the northern border of a site; however, a deeper aquifer may transport contaminants to municipal wells at a much greater distance west of the site. This scenario represents two unique

Figure 6.0.

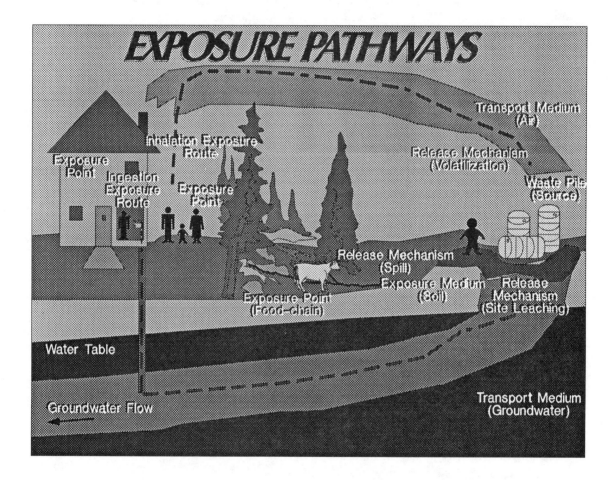

exposure pathways because it involves different aquifers (shallow groundwater versus deep groundwater), different exposure points (residential wells versus municipal wells), and different exposed populations (private well-users versus public well-users). Even though both exposure pathways contain similar elements (source of contamination and exposure route), the link between the elements for the pathways is different. As a result, the two populations may be exposed to significantly different levels of contaminants, and each

exposure pathway may have different health implications.

To determine whether an exposure pathway is relevant to the site, the assessor must have sufficient information to link the contaminated media to a specific population. If information is insufficient to make that determination, the health assessment should show what information is missing and make appropriate recommendations to fill those gaps.

Table 6.0. Exposure Pathways

PATHWAY NAME	EXPOSURE PATHWAY ELEMENTS					TIME
	SOURCE	ENVIRONMENTAL MEDIA	POINT OF EXPOSURE	ROUTE OF EXPOSURE	EXPOSURE POPULATION	

The health assessor should identify the data gaps that exist in each of the five elements, and explain how they affect ATSDR's assessment of a specific pathway. In particular, the health assessor should identify contaminants for which insufficient information exists on transport mechanisms and/or route of exposure and indicate that that data gap prevents complete assessment of the identified contaminant until the data are available. The health assessor has the option of further evaluating this pathway in the Public Health Implications section.

Because of the complexity of identifying exposure pathways for a site area, the health assessor should consider using Table 6.0. as a worksheet to keep track of the five exposure pathway elements.

6.1. IDENTIFYING ELEMENT 1 - SOURCE OF CONTAMINATION

A contaminant source is, as the name implies, the origin of the environmental contamination. Sources of contamination may include the following:

- open burning area;

- detonation area;

- land-spreading area;

- pond;

- lagoon;

- impoundment;

- landfill;

- pit;

- injection well;

- pile;

- material-handling equipment;

- incinerator/boiler;

- tank;

- drum;

- pipe/culvert;

- disposal trench; and/or

- emission stack.

Each source represents a location (point or area) where a release of contaminants to the environment has resulted from some transport mechanism. For example, a land-spreading area is an area where liquids have been poured/leaked into the ground surface or injected or mixed with surface soil. A site may have a single contaminant source or multiple sources (e.g., more than one of these single sources).

The health assessor should recognize the difficulty of identifying the original source of contamination. Although contaminants may have been found in a medium, this does not necessarily mean that the original source of contamination to that medium is known. In many cases, the originating source will not be known. In such cases, the health assessor can consider a contaminated medium as an intermediate source of contamination in order to establish an exposure pathway.

Based on the reports and documents available, the assessor should clearly indicate in the health assessment what is known about the type and extent of contamination at the source and at the receiving media for each exposure pathway. The assessor should state the extent to which contamination sources at the site have been adequately characterized.

Sources of contamination are not always obvious. The health assessor needs to consider a number of factors before deciding whether contamination existed, exists, or could exist in the future. Some of those factors include the following:

- **location or release point;**

- **history of storage, disposal, or release;**

- **contaminants and concentrations at the source;**

- **emission rates ;**

- **frequency of release;**

- **operating period; and**

- **current status.**

In considering the **location or release point** of the source, the health assessor should determine what types of man-made features (e.g., tank, drum, pipe, impoundment, etc.) and natural features (e.g., lake, pond, mineral outcropping, etc.) exist. However, the fact that those features exist does not mean they represent a source of contamination.

History of the site may disclose that contaminants were intentionally or unintentionally disposed of or released at a particular location. If this information has been documented and verified by EPA, the state, the county, or some municipal agency, the disposal area or release point could be considered a source of contamination. The Toxic Chemical Release Inventory (TRI) database also documents various sources and releases of contaminants into local environmental media.

The health assessor should review the **concentration of contaminants** at the suspected release point. Then the assessor should compare them with (1) background concentrations of media samples collected upstream, upwind, uphill, or upgradient, and (2) concentrations of media samples collected downstream, downwind, downhill, or downgradient. This comparison will assist the health assessor in deciding whether the suspected release point should be considered a source of contamination. If the contaminant concentrations decrease with distance downstream, downwind, downhill, or downgradient from a suspected release point or area, and do not increase in the opposite direction, the suspected release point or area may represent a source of contamination.

If no local background concentrations exist for the contaminants of concern, one cannot know whether the concentrations at the release point or area represent (1) the original source of contamination, (2) an intermediate source of contamination, which may not be representative of the original source, or (3) background concentrations, which may be elevated because of local native mineral deposits or widespread industrial or agricultural pollution.

The **emission rate** and **frequency of release** from the source will help determine the importance of evaluating this source further. The **operating period** is crucial in determining the importance of the release point over specific time periods, particularly if the releases involve some human element of control. Knowing the operating period and the **current status** will assist in determining the relevant time period that should be considered when assessing the exposure pathway.

6.2. IDENTIFYING ELEMENT 2 - ENVIRONMENTAL MEDIA AND TRANSPORT

After identifying the contaminant source, the assessor must identify all the environmental media that may serve to transport contaminants from this source to possible points of human exposure (see Section 6.3.). Affected environmental media may include waste materials, leachate, soil gas, sludge, surface soil, subsurface soil, sediment, surface water, groundwater (with subcategories of private wells, public wells/systems, and monitoring wells), air, and biota.

Once the contaminated media have been identified using procedures in Chapter 5, the assessor must consider the transport and transformation mechanisms that could influence contaminant migration via those media. The assessor should first focus on the

sampled media that are known to be contaminated. The sampling date of those media, the date of any remedial measures on the site, and any other date when site actions may have affected media concentrations should be reviewed.

When more than one environmental medium may be involved in transport of the contaminant from the source to the receptor population, the assessor should explain that in the narrative. It is important for the health assessor to know both the past and present status of contamination for these media. When media have not been adequately sampled, the assessor should explore the fate and transport processes to determine whether the media have been, are now, or may become contaminated in the future.

Once the transport processes for a medium have been assessed, and the possible extent of contamination from the source has been determined, the assessor should be ready to determine the point or points of exposure associated with that contaminated medium. When using Table 6.0., the assessor should state only the environmental medium that occurs at the point of exposure (Section 6.3.), not all media that may have been involved in inter-media transfer (e.g., contaminants moving from one medium to another).

6.2.1. Fate and Transport Mechanisms

The evaluation of transport mechanisms is very important in determining the following elements:

- the original source of the contamination and the release point for an exposure pathway;

- the likelihood of contamination and potential exposure beyond the sampled areas;

- the representativeness and adequacy of the environmental sampling conducted at the site (see Section 6.2.4.);

- the need and urgency for additional environmental sampling, exposure assessment, or other health activity; and/or

- estimates of the length of time that environmental media and points of exposure may have been contaminated.

In general, environmental transport involves the movement of gases, liquids, and particulate solids within a given medium and across interfaces between air, water, sediment, soil, plants, and animals. Once a substance is released to the environment, one or more of the following may occur:

- movement (e.g., advection/convection in water or transportation on suspended sediment or through the atmosphere);

- physical transformation (e.g., volatilization, rain);

- chemical transformation (e.g., photolysis, hydrolysis, oxidation/reduction, etc.);

- biologic transformation (e.g., biodegradation); and/or

- accumulation in one or more media (including the medium receiving the release).

Transport and fate mechanisms can usually be simplified into four basic categories:

1. emission (the actual release or discharge of the contaminated material from a source);

2. advection or convection (the normal migration or movement of the contaminant through a medium, e.g., stream flow, air flow, surface runoff, soil erosion, soil creep, mass movement, etc.);

3. dispersion (spreading of contaminants in a liquid, gas, or solid phase due to impingement of the contaminant by that phase material); and

4. attenuation (the retardation, degradation or adsorption of a contaminant).

Examples of emission, advection, dispersion, and attenuation for soil and surface water are shown in Table 6.2.1., along with the inter-media transfers that might occur. In addition, Figures 6.2.1.a. and 6.2.1.b. illustrate two examples of how those four transport categories apply to a discharge into a waterway and to stack emissions. For further explanation of the terms used in Table 6.2.1., please refer to the definitions in section 6.2.2. During the analyses of contaminant fate and transport in an environmental medium, the health assessor should try to answer the following questions:

- At what rate are contaminants entering the medium? (emission rates)

- Where are the contaminants going and how fast are they migrating? (advection)

- How are the contaminants spreading out in the medium? (dispersion)

- What is the degree of buffering or degradation of contaminants as they migrate? (attenuation)

- Will contaminants migrate to another medium? (inter-media transfers)

- What about past and future exposures?

For each contaminated environmental medium present on- or off-site, there may be several transport mechanisms that should be considered. For example, where contaminated soil exists, the assessor should try to determine whether the contaminants of concern are transported via surface-water runoff, leaching, volatilization, airborne suspension and resuspension, or biologic uptake. Tables 6.2.2. and 6.2.3. depict important transport mechanisms available for environmental media that may be represented on- or off-site. Refer to sections 6.2.2. and 6.2.3. for a discussion of the terms used in the two tables.

6.2.2. Chemical-Specific Factors Influencing Environmental Fate and Transport

After identifying the contaminants of concern in the sampled environmental media and the

Table 6.2.1. Transport Mechanisms

TRANSPORT CATEGORY	MEDIUM	
	SOIL	SURFACE WATER
EMISSION (Affecting Medium of Interest)	• Mass of contaminant loading per day (e.g., spill/release from drum, pipe, or truck)	• Pipe discharges or spills • Surface runoff and soil loss • Deposition from air • Groundwater discharges
ADVECTION	• Infiltration • Soil gas migration • Soil creep • Erosion via wind or water	• Stream flow • Lake currents and turnovers
DISPERSION	• Impingement with soil particles	• Mixing zone in watercourse or water body
ATTENUATION	• Adsorption • Biodegradation • Hydrolysis • Oxidation/reduction • Photolysis • Volatilization	• Sedimentation and the others listed for soil
INTER-MEDIA TRANSFER	• Migration of gases or particles to air • Migration to groundwater • Migration to surface water, sediments, and aquatic biota • Biologic uptake into plants and animals	• Sediment adsorption • Bioaccumulation • Gas migration to air • Recharge into groundwater

Figure 6.2.1.a.

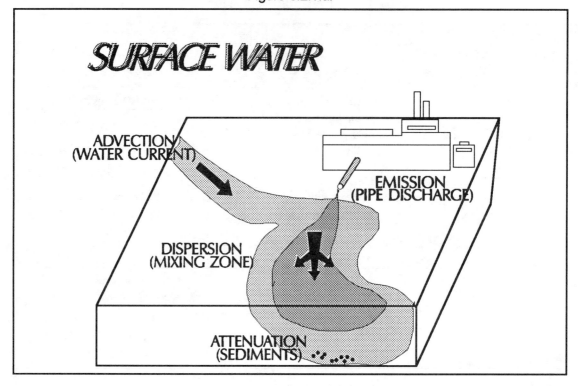

Figure 6.2.1.b.

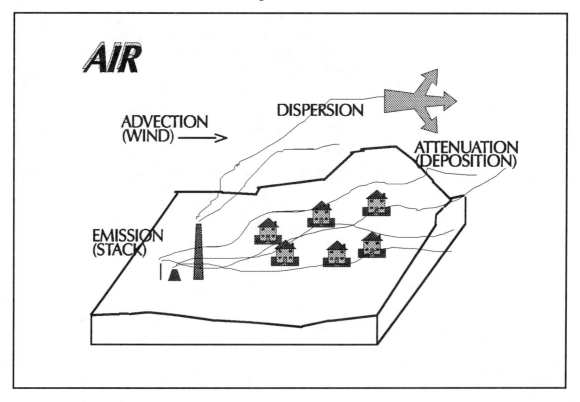

possible transport mechanisms (see chapter 5), the assessor should consider chemical-specific factors that may influence contaminant transport. Table 6.2.2. contains a list of some important chemical- and site-specific factors that may affect transport. Physicochemical properties, such as solubility and density, influence a chemical's fate and transport across interfaces and through environmental media. Some chemical-specific factors are discussed in the following paragraphs.

Water solubility refers to the maximum concentration of a chemical that dissolves in a given amount of pure water and usually ranges from 1-100,000 mg/L. The solubility is very important when it comes to understanding a contaminant's ability to migrate in the environment. Environmental conditions, such as temperature and pH, can influence chemical solubility. Highly water-soluble chemicals are less strongly adsorbed to soil and, thus, are rapidly leached from contaminated soil into both groundwater and surface water. Solubility also affects volatilization from water. For example, highly water-soluble chemicals tend to be less volatile and are also more readily biodegradable (1). For liquids that are immiscible with water, liquid density plays a critical role. Liquids that are denser than water may penetrate and preferentially settle to the base of the aquifer, while lighter liquids will float. The water solubility of numerous chemicals can be found in a variety of sources, such as ATSDR Toxicological Profiles, the *Hazardous Chemicals Desk Reference* (2) or EPA's *Integrated Risk Information System (IRIS)* database.

Vapor pressure is a measure of the volatility of a chemical in its pure state and is an important determinant of the rate of volatilization from contaminated surface soils and water bodies to air. The temperature, wind speed, and soil conditions of a particular site, as well as the adsorption characteristics and water solubility of the chemical, will affect volatilization rates. In general, chemicals with relatively low vapor pressures and a high affinity for soil or water are less likely to vaporize and become airborne than chemicals with a high vapor pressure and less affinity for soil or water. The vapor pressure can be found for numerous chemicals in a variety of sources including ATSDR's Toxicological Profiles and EPA's IRIS database.

Henry's law constant (H) takes into account molecular weight, solubility, and vapor pressure, and indicates the degree of volatility of a chemical in a solution (3). When the chemical contaminant has high water solubility

Henry's Law Constant Ranges

Extent of Volatility	Range of Values (atm m^3/mol)
nonvolatile	less than 3×10^{-7}
low volatility	3×10^{-7} to 1×10^{-5}
moderate volatility	1×10^{-5} to 1×10^{-3}
high volatility	greater than 1×10^{-3}

Table 6.2.2. Chemical-Specific Factors Affecting Transport Mechanisms
(continues)

Medium	Transport Mechanisms	Factors Affecting Transport
Groundwater	• Groundwater movement (via hydrolic connection between aquifers or with surface water and man-made objects, e.g., improperly cased wells, sewers, and conduits)	• Density • Water solubility • K_{oc} (organic carbon partition coefficient)
	• Volatilization	• Water solubility • Vapor pressure • Henry's Law Constant
	• Adsorption to soil particles or precipitation out of solution	• Water solubility • K_{ow} (octanol/water partition coefficient) • K_{oc}
	• Biologic uptake	• K_{ow}
Soil (Surface and Subsurface) Sediment Sludge Waste Materials	• Runoff (soil erosion)	• Water solubility • Koc
	• Leaching	• Water solubility • Koc
	• Volatilization	• Vapor pressure • Henry's law constant
	• Biologic uptake	• Bioconcentration factor

in relation to its vapor pressure, the chemical dissolves mainly in water. When vapor pressure is high relative to water solubility, Henry's law constant is high and the chemical volatilizes primarily to the air. A high Henry's law constant for a pollutant would suggest inhalation as a potential route of exposure. Henry's law constant can be found for a number of chemicals in ATSDR's Toxicological Profiles.

Table 6.2.2. (continued)

Medium	Transport Mechanisms	Water Solubility
Surface Water	• Overland flow (via natural drainage or man-made channels)	• Water solubility • K_{oc}
	• Volatilization	• Water solubility • Vapor pressure • Henry's law constant
	• Hydrologic connection between surface water and groundwater	• Density
	• Adsorption to soil particles	• Water solubility • K_{ow} • K_{oc}
	• Sedimentation (of suspended and precipitated particles)	• Density • Water solubility
	• Biologic uptake	• K_{ow} • Bioconcentration factor
Air	• Aerosolization	• Water solubility
	• Atmospheric deposition	• Particle size
	• Volatilization	• Henry's law constant
Biota	• Bioaccumulation	• K_{ow}
	• Bioconcentration	• Bioconcentration

The K_{oc} (**organic carbon partition coefficient**) (also known as the soil/water partition coefficient or the adsorption coefficient) is a measure of the tendency for organic compounds to be adsorbed by soil and sediment. The K_{oc} is chemical-specific and largely independent of soil properties. K_{oc} values range from 1 to 10^7. A high K_{oc} indicates that organic chemicals bond tightly to organic matter in the soil so that less of the chemical is available to move into groundwater or surface water. A low K_{oc} suggests the potential for chemical movement into groundwater or surface water. A K_{oc} can be found for a number of chemicals in ATSDR's Toxicological Profiles.

K$_{oc}$ Ranges (ml/g organic carbon)

Sorption to Soil	Coefficient Values
very weakly sorbed	less than 10
weakly sorbed	10 to 100
moderately sorbed	100 to 1,000
moderately to strongly sorbed	1,000 to 10,000
strongly sorbed	10,000 to 100,000
very strongly sorbed	greater than 100,000

The K$_{ow}$ (octanol/water partition coefficient) predicts the chemical's potential for accumulation in animal fat by measuring how a chemical is distributed at equilibrium between octanol and water. Organisms tend to accumulate chemicals with high K$_{ow}$ constants in the lipid portions of their tissues. Thus, one way to estimate the bioconcentration potential of a chemical is to measure how lipophilic it is. Because it is difficult to measure a chemical's lipophilicity directly, researchers typically use the K$_{ow}$ value to predict a chemical's tendency to partition between an octanol component (a fat surrogate) and water. It is directly related to a chemical's tendency to bioconcentrate in biota and inversely correlated with water solubility. Chemicals with large K$_{ow}$ values tend to accumulate in soil, sediment, and biota. For example, lipophilic compounds, such as dioxin, DDT, and polychlorinated biphenyls (PCBs), are soluble in lipophilic matter. This class of chemicals tends to bioaccumulate in biota; adsorb strongly onto soil, sediment, and organic matter; and transfer to humans through the food chain. Conversely, chemicals with small K$_{ow}$ values tend to partition mostly into air or water. For example, volatile organic compounds (VOCs), such as trichloroethylene and tetrachloroethylene, tend to be widely distributed in air, and exposure via the food chain is of less concern than other exposure pathways.

The uptake of soil contaminants by plants has not been systematically studied. Some chemicals, such as cadmium, are readily taken up from soil by growing plants (4). For other contaminants, such as polychlorinated biphenyls (PCBs), it appears that contaminant volatilization from soil and deposition on plant surfaces may be more important than uptake by the root and translocation (5).

A K$_{ow}$ can be found for a number of chemicals in ATSDR's Toxicological Profiles.

The **bioconcentration factor** (BCF) is a measure of the extent of chemical partitioning at equilibrium between a biologic medium, such as a fish tissue, and an external medium, such as water. A bioconcentration factor is determined by dividing the equilibrium concentration (mg/kg) of a chemical in an organism or tissue by the chemical concentration in the external medium. In general, chemicals with a high K$_{ow}$ value tend to have high BCFs (6). However, some compounds such as polyaromatic hydrocarbons (PAHs), do not significantly bioconcentrate in fish and vertebrates in spite of having a high K$_{ow}$ value. The relative absence of PAH bioconcentration in fish is due to the fish's ability to rapidly metabolize PAH compounds (7).

Bioaccumulation is a broader term, which refers to a process that includes bioconcentration and any uptake of chemical residues from dietary sources. A bioconcentration factor can be found for a number of chemicals in EPA's IRIS database, the National Library of Medicine's TOXNET Hazardous Substances Data Bank, and in ATSDR's Toxicological Profiles.

Transformation and degradation rates take into account physical, chemical, and biologic changes in a contaminant over time. Chemical transformation is influenced by hydrolysis, oxidation, photolysis, and microbial degradation. A key transformation process for organic pollutants is aqueous photolysis, i.e., the alteration of a chemical species due to the absorption of light. Biodegradation, the breakdown of organic compounds, is a significant environmental process in soil. The rate of biodegradation is a function of the organic content of the soil. Precise estimations of chemical-specific transformation and degradation rates are difficult to calculate and to apply because they are subject to site-specific physical and biologic variables. For more information, refer to ATSDR's Toxicological Profiles for the section on environmental fate, transport, partitioning, transformation and degradation of chemicals in air, water, and soil. In addition, the National Library of Medicine's TOXNET Hazardous Substances Data Bank has data fields for environmental fate and transformations.

Knowledge of the factors discussed here will assist the assessor in understanding the chemical's behavior in the environment and will help to focus the assessment on transport mechanisms of the greatest significance. This information should not be used to justify lack of media-specific contaminant data; rather, it should be used to assist the assessor in identifying transport mechanisms that may lead to human exposure.

6.2.3. Site-Specific Factors Influencing Environmental Fate and Transport

When identifying possible transport pathways, the assessor should also consider site-specific factors that may affect contaminant transport. Table 6.2.3. lists some of those important factors. Each site is unique and must be evaluated to determine characteristics that could enhance or retard the migration of contaminants of concern. Many factors affecting transport depend on climatic conditions and physical characteristics of the site. The assessor should evaluate site-specific factors in view of the characteristics of the contaminants of concern to gain an understanding of their migration potential. Site-specific factors that should be considered are discussed in the following paragraphs.

Annual precipitation rates may be useful in determining the amount of surface-water runoff, groundwater recharge rates, and soil moisture content. If high precipitation rates are coupled with highly water-soluble contaminants of concern, the extent of contaminant migration may be great. In addition, precipitation promotes the scavenging of particulates and soluble vapors from the atmosphere.

Temperature conditions affect the volatilization rate of contaminants. In addition, ground temperature can affect the movement of contaminants, e.g., frozen ground retards movement.

Wind speed and direction influence generation rates of fugitive dust. During periods of atmospheric stability, gravitational settling will act to redeposit suspended particulates or droplets.

Seasonal and diurnal conditions could be a major factor affecting rates of contaminant migration where precipitation rates or temperatures vary greatly according to the season or time of day.

Geomorphologic characteristics of the site play a significant role in determining stream flow velocity, volume and speed of runoff, erosion rates, and soil characteristics. Karst terrains developed on limestone bedrock can significantly enhance hydrologic connection between surface water and groundwater.

Hydrogeologic characteristics (types and locations of aquifers, i.e., water table or confined, and hydraulic conductivity) are important in determining the threat the site may pose to drinking water supplies. The assessor should use generalized geologic profiles cautiously and should use information from site-specific well tests to evaluate the connection between aquifers and the continuity of aquitards. Water table contours and piezometric surfaces indicate hydraulic gradients and resulting groundwater flow patterns, including the potential for surface discharges (i.e., seeps, springs, and influent streams).

Table 6.2.3. Site-Specific Factors Affecting Transport Mechanisms (Continues)

Medium	Transport Mechanisms	Factors Affecting Transport
Groundwater	• Groundwater movement (via hydrologic connection between aquifers or with surface water and man-made objects, e.g., improperly cased wells, sewers, and conduits)	• Site Hydrogeology (karst, alluvial deposits, fractures, continuity of aquitards) • Precipitation • Infiltration rate • Groundwater direction • Depth to aquifer • Influent and effluent streams • Presence of other compounds • Soil type • Soil chemistry • Presence and condition of wells (well location, depth, use, casing material, and construction) • Conduits, sewers
	• Volatilization	• Depth to water table • Soil type and cover • Climatologic conditions • Contaminant concentration
	• Adsorption to soil particles or precipitation out of solution	• Presence of other compounds • Soil type and chemistry • Presence of other compounds
	• Biologic uptake	• Groundwater use for irrigation and livestock watering

Table 6.2.3. (continued)

Medium	Transport Mechanisms	Factors Affecting Transport
Surface Water	• Overland flow (via natural drainage or man-made channels)	• Precipitation (frequency, duration) • Infiltration rate • Vegetative cover and land use • Soil type and chemistry • Use as potable water • Location, width, and depth of channel, velocity, dilution factors, direction of flow • Floodplains • Point and nonpoint source discharge areas
	• Volatilization	• Climatologic conditions • Surface area • Contaminant concentration
	• Hydrologic connection between surface and groundwater	• Influent and effluent streams • Stream bed permeability • Soil type and chemistry
	• Adsorption to soil particles	• Particle size and density
	• Sedimentation (of suspended and precipitated particles)	• Particle size and density
	• Biologic uptake	• Chemical concentration • Presence of plants and animals

Table 6.2.3. (continued)

Medium	Transport Mechanisms	Factors Affecting Transport
Soil, Sediment, Sludge	• Runoff (soil erosion)	• Presence of plants • Soil type and chemistry • Precipitation rate • Configuration of land and surface condition
	• Leaching	• Soil type • Soil porosity and permeability • Soil pH • Cation exchange capacity • Organic carbon content
	• Volatilization	• Physical properties • Chemical properties
	• Suspension and resuspension	• Climate • Presence of plants • Site activities and traffic • Rainfall
	• Biologic uptake	• Soil properties • Contaminant concentration

Table 6.2.3. (continued)

Medium	Transport Mechanisms	Factors Affecting Transport
Air	• Wind	• Speed, direction, atmospheric stability
	• Aerosolization	• Chemicals stored under pressure
	• Atmospheric deposition	• Rainfall
Biota	• Biomagnification	• Presence of plants and animals • Consumption rate
	• Migration	• Commercial activities (farming, aquaculture, livestock, dairies) • Sport activities (hunting, fishing) • Migratory species
	• Vapor sorption	• Soil type • Plant species
	• Root uptake	• Contaminant depth • Soil moisture • Plant species
Waste Materials (e.g., exposed wastes, mine tailings, waste liquids, drums)	• Surface water runoff • Leaching • Groundwater movement • Volatilization	• All mechanisms: Waste type Integrity of contaminant Climatic conditions

Surface-water channels (location, width, and depth) and associated floodplains near the site may also affect the extent of contaminant migration. Variations of flow with seasonal changes or intermittent streams should be noted. Because effluent streams receive water from the zone of saturation (i.e., the channel lies below the water table), they can enhance contaminant movement from contaminated groundwater to surface water. Because influent streams (i.e., the channel lies above the water table) contribute water to the zone of saturation, they can enhance contaminant movement from surface water into groundwater.

Soil characteristics (including the configuration, composition, porosity, permeability, and cation exchange capacity of the soil) influence rates of percolation, groundwater recharge, contaminant release, and transport. Information on background levels of metals and organic compounds and on pH levels in area soils is needed to ensure the extent of contamination has been delineated.

Ground cover and vegetative characteristics of the site influence rates of soil erosion, percolation, and evaporation.

Plants and animals at or near the site could be used for human consumption and can enhance the rate of transport to receptors.

Man-made objects, such as sewers, culverts, and drainage channels, can increase the movement of contaminants. Improperly constructed wells can lead to interaquifer contamination.

For information on the climatic factors discussed, refer to the National Oceanic and Atmospheric Administration's database (Local Climatological Data Annual Summaries, parts I-V, available from RPB/ESS or the National Climatic Data Center, Asheville, NC). For information on physical and chemical characteristics of waterways and lakes, geomorphic characteristics, and hydrologic characteristics of groundwaters and surface waters, contact the U.S. Geological Survey, the

State Geological Survey, or the local county or municipal engineer. For information on federal navigation channels, flood control reservoirs, and sediment quality in waterways and lakes, contact the U.S. Army Corps of Engineers. For information on public water supplies, contact EPA, the state health department or environmental agency, or the local public health department. For information on soils and ground cover, contact the Soil Conservation Service of the U.S. Department of Agriculture. For information on plants and animals, contact local game wardens, state game management and fish departments, the County Extension Agent or Agricultural Agent of the U.S. Department of Agriculture, and/or the U.S. Fish and Wildlife Survey.

6.2.4. Representativeness and adequacy of the environmental sampling conducted at the site.

This section provides guidance for determining whether environmental data are sufficient to characterize contaminant transport and the extent of contamination. Section 5.3.3. provides additional information on this topic for each of the media. The health assessor should conduct the following steps:

1. Review the deficiencies in the number, location, time coverage, and quality of the samples.

2. Mention explicitly those media that have not been sampled.

3. Consider each of the following recommendations when evaluating the specific medium of concern.

SURFACE WATER

• Review the site visit report or inspect a USGS topographic map for slope information and stream flow direction.

• Determine the location of stations (e.g., upstream, downstream, downcurrent, etc.) relative to the site and duration of sampling to assess representativeness of data.

- Understand that short-term sampling detects only what exists in the surface water at a particular time; it provides poor data on contaminant transport.

- Note that contaminant concentrations in surface waters of both waterways and water bodies can fluctuate significantly because of meteorologic conditions, local geohydrology, thermal stratification, and seasonal events.

GROUNDWATER

- Verify whether sufficient information exists on local private and public wells, the location and populations using these wells, type of well use, well depth, and length of use. Use this information to help determine whether sampling data are representative of the extent of groundwater contamination.

- If the groundwater flow direction cannot be obtained from EPA or site documents, contact the U.S. Geological Survey and/or the State Geological Survey for local hydrogeologic information.

- Consider using water table elevations of monitoring wells and other wells in the vicinity to get some idea of the direction of groundwater flow in the water table aquifer.

- Determine whether groundwater samples were unfiltered and were collected from upgradient and downgradient areas and from the site's groundwater contamination plume.

AIR

- When assessing air data, state how useful the data are based upon sampling duration. Indicate if the sampling duration is representative of acute (less than 14 days), intermediate (15-364

days), or chronic (greater than 365 days) exposure.

- Indicate whether or not the air sampling was conducted at the breathing zone.

- Verify, using on-site wind rose information, that the air monitoring station(s) is (are) located downwind from the site. If on-site meteorologic data do not exist, state the distance to the closest meteorologic station with wind rose data and discuss how representative it is of the site.

- Explain whether windborne dust was observed coming from the site areas and whether dust tracks or deposits on the ground surface or snow were observed.

- For information on soil composition, porosity, and permeability, contact the soil conservation service to find out how soil in the area will influence rates of percolation and possible contaminant transport. When possible, consider examining the detailed soil survey that exists.

- Summarize any biases that may be associated with using soil data. If soil samples were collected as 1) grabs from a single location or depth; 2) composites from a single location, single depth in a core, or multiple depths in a core; and 3) composites from multiple locations or multiple cores, explain how the sampling method could affect the data.

SEDIMENT AND SLUDGE

- If samples did not include cores of sludge deposits on the site or cores of sediments in depositional zones, note that the assessment may not be able to address the possibility of long-term releases from those areas.

BIOTA

- If contaminant uptake in fish represents a possible transport mechanism, be sure that there is evidence (fish data) that

Figure 6.2.4. (Continues)

FLOW CHART FOR FATE AND TRANSPORT ASSESSMENT

Environmental fate and transport assessment: atmosphere

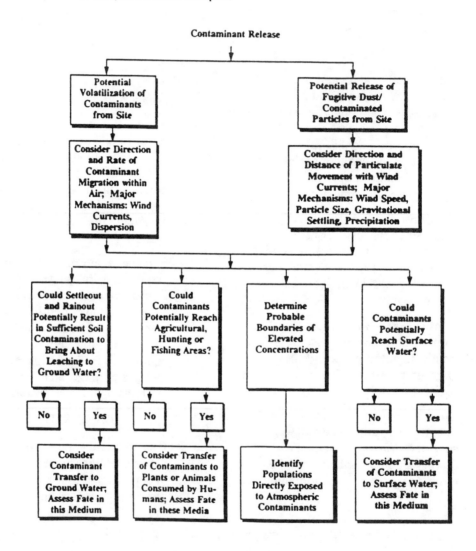

Source: Adapted from EPA 1989. [11]

Figure 6.2.4. (continued)

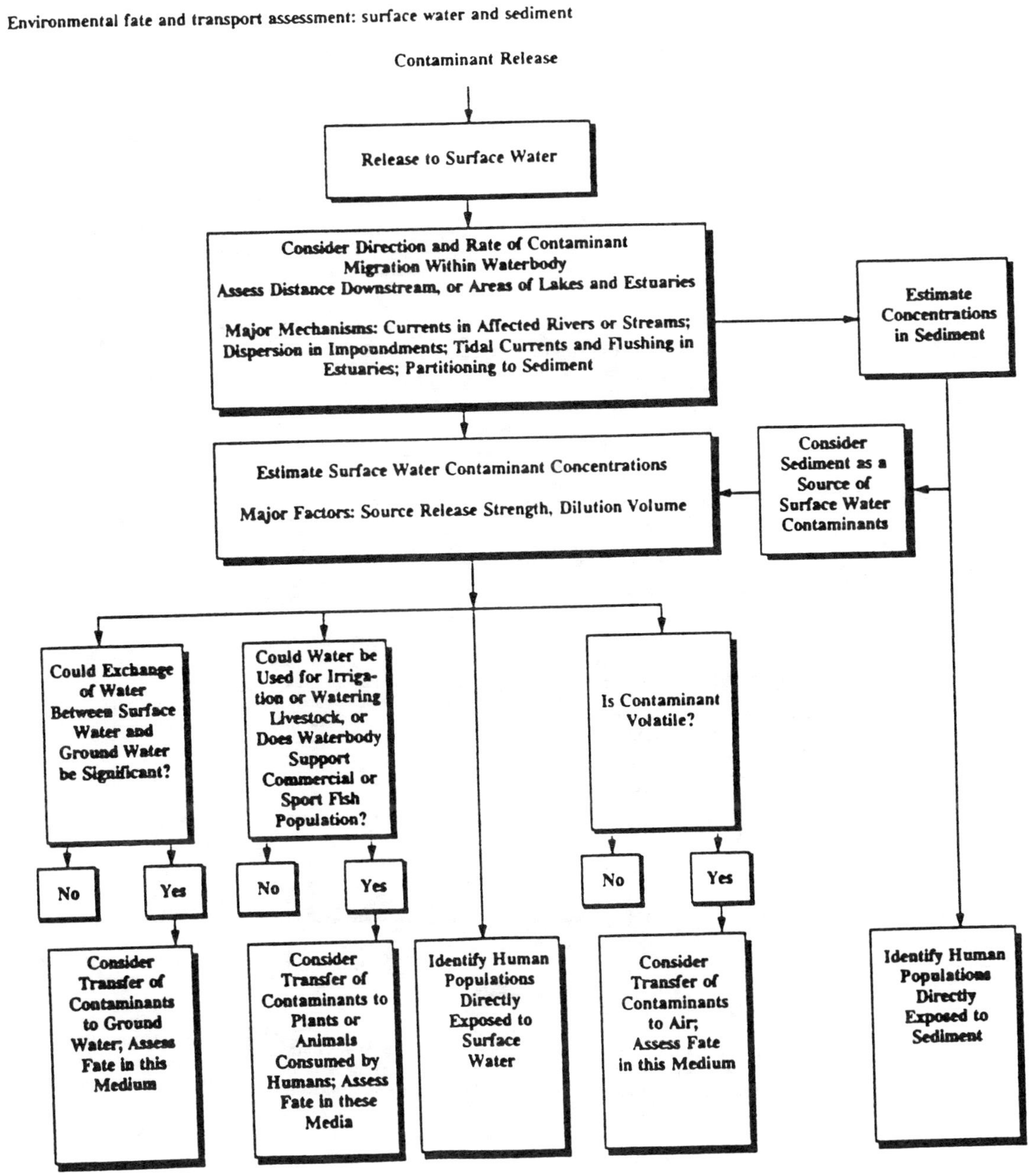

Environmental fate and transport assessment: surface water and sediment

Source: *Adapted from EPA 1989.* [11]

Figure 6.2.4. (continued)

Environmental fate and transport assessment: soils and ground water

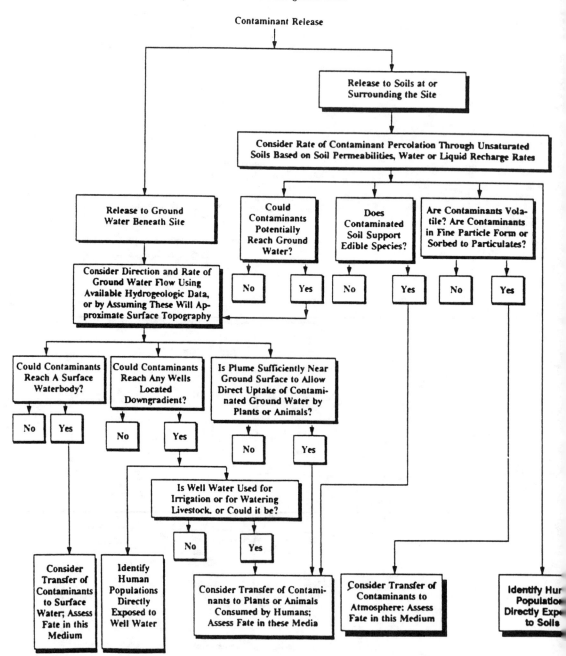

Source: Adapted from EPA 1989.[11]

local fish are affected by site contaminants. If not, determine whether sufficient data exist on the environs (surface waters and sediments) that suggest possible uptake in fish, and explain why they may be affected by certain site contaminants.

If local biota are contaminated by an environmental medium, determine if any research shows whether contaminant uptake at the levels found could occur. Cite this research.

If data for a specific environmental medium are sufficient, summarize the contaminants and the extent, rate, and direction of migration on and off the site in that medium and how that migration relates to the exposure pathway of concern.

To assist the health assessor in evaluating the fate and transport of contaminants in air, surface water, sediment, soils, and groundwater, included here are three flow charts that EPA developed in its *Risk Assessment Guidance for Superfund, Volume I, Human Health Evaluation Manual (Part A) Interim Final* (11). The use of these flow charts (Figure 6.2.4.) and/or some variation of them should encourage a more consistent approach in assessing the fate and transport of contaminants.

6.2.5. Environmental Transport Models

Environmental modeling is used for both qualitative and quantitative analysis of contaminant transport at a site. There are some instances when the assessor may use environmental transport models to assist in conceptualizing the transport mechanisms and the factors affecting them at a site. Some specific uses of environmental modeling include the following:

- to evaluate and recommend sampling locations;

- to identify data and information gaps;

- to describe temporal and spatial trends for contaminant concentrations at an exposure point;

- to estimate the duration (e.g., years) of exposure; and

- to estimate contaminant concentrations at exposure points in the past or in the future when monitoring data are not available.

The assessor should recognize that models contain assumptions that influence the validity of the predictions derived from the model. Therefore, before relying on data derived from modeling, the assessor should ensure, whenever possible, that the model being used has been validated and calibrated using site-specific data. The validity of those assumptions is often difficult to establish on a site-specific basis. For that reason, ATSDR believes that such models or mathematic expressions are tools that should be used only as guides to help develop public health decisions and should not replace decisions based on comprehensive, validated, environmental samples. The health assessor should not use the predicted environmental concentrations from models to estimate exposure doses in humans, or to draw health conclusions. ATSDR policy states that modeling cannot serve as a proxy for actual measurements of existing conditions when determining public health implications; that fact should be noted if the assessor uses modeling data.

Although ATSDR strongly recommends the use of validated analytical data as the basis for public health conclusions, it may be necessary to rely on modeling data when analytical data are not available. If modeling data are used in the health assessment, the data should be clearly identified as such, and the uncertainty and limitations of the data should be explicitly acknowledged. Where modeling is performed, the assumptions used in each model should be stated. Whenever feasible, the health assessment should include a recommendation to obtain additional environmental analytical data to confirm any conclusions based on modeling data. Additional information on the use of environmental transport models is presented in the EPA *Superfund Exposure Assessment Manual* and the EPA *Risk Assessment Guidance for Superfund* (10,11).

6.3. IDENTIFYING ELEMENT 3 - POINT OF EXPOSURE

The point of exposure, the third element of an exposure pathway, is the point at which people contact a contaminated medium. It can be identified by reviewing past, present, and possible future use of land and natural resources. Historically, patterns of urban land use have varied widely. A site may have served a number of uses (e.g., recreational, residential, agricultural, commercial, and industrial) that resulted in a variety of exposure points, depending on the contaminated media and specific timeframe examined. Therefore, it is important that the assessor consider past, present, and future land and natural resource use. Because of remedial measures or other site-related activities, the health assessor may not find any existing exposure points. However, the health assessor should recognize that past exposure points existed and try to identify them.

Where the presence of physical controls and barriers (e.g., permanent fences, gates, etc.) or institutional controls (e.g., ordinances, building permits, etc.) prevent contact with the contaminated medium of concern, health assessors should assume that no exposure point exists for persons unable to gain access to the contaminated medium. Despite these controls, some persons (e.g., trespassers, remedial workers, etc.) may still have access to the contaminated medium. Where evidence exists that persons had or still have access to the contaminated medium, the health assessor should state that an exposure point existed or exists.

Possible exposure points for each environmental medium are discussed below.

Groundwater exposure points include wells used for municipal, domestic, industrial, and agricultural purposes. Groundwater may also be used as a water supply source for swimming pools and other recreational water activities. In some areas, natural springs that are groundwater discharge areas are used for recreation.

Surface-water exposure points include irrigation and public, industrial, and livestock water supplies.

Soil may serve as an exposure point for on-site workers. Contaminated subsurface soils may serve as an exposure point for workers involved in excavation or drilling. Contaminated off-site surface soil in residential areas is an exposure point for area residents. Indoor exposure points are also of concern; these may result from off-site transport to residences via suspension, deposition, and tracking.

Air exposure points involve contaminants that are volatile or adsorbed to airborne particulates and may occur indoors or outdoors. Structures surrounding a site may be an exposure point for indoor airborne contaminants from migrating soil gases. The area downwind of a site might be an exposure point for contaminated ambient air as a result of volatilization or entrainment of contaminants in dust particles. A wind rose may be useful in evaluating the transport of airborne contaminants in various directions from a site.

Food-chain exposure points occur if people consume plants, animals, or other food products that have contacted contaminated soil, sediment, waste materials, groundwater, surface water, air, or biota. Therefore, garden and orchard produce, nut-tree products, irrigated crops, livestock, game, medicinal plants, and other aquatic and terrestrial organisms at or near the site may act as exposure points for humans. The assessor should also consider exposure points that may occur when contaminated materials are used as livestock feed.

Other media may provide additional exposure points. Sediments can serve as contaminant reservoirs for aquatic organisms or can be transported to other areas and used as top soils. Contaminated building materials at commercial or industrial sites may be removed and used off-site. Sludges from industrial or municipal waste-treatment processes and dredged materials may also be points of exposure.

After identifying the exposure points at a site, the assessor must consider the potential routes of exposure at the exposure point.

6.4. IDENTIFYING ELEMENT 4 - ROUTE OF EXPOSURE

The exposure route is the fourth element of an exposure pathway. Exposure routes are the means by which contaminants enter the human body. They include:

- ingestion of contaminants in groundwater, surface water, soil, and food;

- inhalation of contaminants in groundwater or surface water via steams and aerosols, air, or soil;

- dermal contact with contaminants in water, soil, air, food, and other media; and

- dermal absorption of contaminants in water, soil, air, food, and other media.

The assessor should consider all exposure routes relevant to the site, including infrequently considered routes, but focusing on the more probable exposure routes. Table 6.4. lists major exposure routes that should be considered for each environmental medium in determining the likelihood of human exposure. Where the presence of physical controls and

barriers (e.g., permanent fences, gates, etc.) or institutional controls (e.g., ordinances, building permits, etc.) prevent contact with the contaminated medium of concern, health assessors should assume that no exposure route exists. Despite those controls, some persons (e.g., trespassers, remedial workers, etc.) may have access to the exposure point. Where evidence exists that persons do have access, that fact should be documented.

If the route of exposure is likely to be incidental for all persons, the health assessor should state that and determine whether the exposure pathway should be dropped from further consideration. One potential route of exposure that should be addressed is the possibility of pica ingestion by pica persons with access to residential yard soil or other contaminated materials.

Once an exposure route is determined relevant for an exposure pathway, the duration and frequency of the exposure route should be discussed for the pathway. Each exposure timeframe, when relevant to the site, must be addressed in the Public Health Implications section (Chapter 7).

6.5. IDENTIFYING ELEMENT 5 - RECEPTOR POPULATIONS

The final element of an exposure pathway is a population that is exposed or potentially exposed through the identified exposure routes to contaminants at an exposure point. Each exposure pathway should be considered with respect to populations (e.g., workers, remedial workers, residents, transient populations, hunters, and fishermen) that may contact that pathway. Exposed populations should be identified as accurately as possible. For example, if the only exposure pathway is via contaminated soil in a residential area along the northern border of a site, the residents of houses in that area are the population of concern for that particular pathway, not all residents living within a one-mile radius of the site. However, all users of a municipal water supply would constitute the exposed population of concern if a municipal well was shown to be contaminated. If private wells are shown to be contaminated, then the currently exposed population would be only the users of those private wells.

If more than one time component exists, the health assessor should try to provide a population estimate for each one. When populations cannot be identified in conjunction with an exposure pathway in the past, present, or future, that exposure pathway is not relevant to the site and should be so noted in the health assessment. The assessor should review data resources for information on the locations of populations and factors affecting their exposure.

6.5.1. Location of Populations

The assessor should determine the location of persons at or near the site. Homes, schools, retirement centers, parks, other recreational areas, and transportation routes should be noted. Proximity to the site can indicate the magnitude and frequency of contaminant exposure. Some locations, such as beaches, tourist attractions, hotels, and other establishments along major transportation routes, should be noted because they involve transient populations that may be exposed during their stay in the area. Populations that may be exposed to contaminants can be identified by considering exposure routes at each exposure point.

Populations exposed via contact with water. Potentially exposed persons include those who are using contaminated water for bathing or recreation. Where possible, populations using water from sources subject to contamination should be identified and demographic characteristics noted that may affect the amount of exposure.

Populations exposed via inhalation. Because contaminant concentrations in the atmosphere will vary based on the characteristics of the release and the distance from the pollutant source, affected populations may contain persons exposed to varying degrees of the contaminant. Isopleth maps of contaminant concentrations (contour maps representing areas of contaminant isoconcentrations) can be overlain on census maps to identify populations exposed to particular levels of the contaminant in the atmosphere (8).

Populations exposed via ingestion of soil. The population exposed to contaminants through soil ingestion can be identified by estimating the area of contaminant dispersion and then determining the population within the contaminated area. The population may be further characterized by identifying individuals who are more likely to ingest soil (i.e., children) (9). However, the entire population in the contaminated area may ingest some soil.

Populations exposed via ingestion of drinking water. Identifying the population

Table 6.4. Media-Specific Exposure Routes

Medium	Exposure Routes
Water	1. Direct ingestion 2. Dermal contact and reaction Ocular contact and reaction 3. Inhalation secondary to household use
Soil	1. Direct ingestion (primarily by children 9 months to 5 years of age) 2. Dermal contact and reaction Ocular contact and reaction 3. Inhalation of chemicals volatilized from soil 4. Inhalation of reentrained dust
Air	1. Inhalation 2. Dermal contact and reaction Ocular contact and reaction
Biota/Food Chain	1. Ingestion of plants, animals, or products contaminated secondary to intake of contaminated water 2. Ingestion of plants, animals, or products contaminated secondary to intake of or contact with contaminated soil, dust, and air 3. Ingestion of plants, animals, or products contaminated secondary to inhalation or evapotranspiration of contaminated air 4. Dermal contact with and reaction to contaminated plants, animals, or products
Miscellaneous Media	1. Direct ingestion 2. Dermal contact and reaction; ocular contact and reaction 3. Inhalation secondary to volatilization or reentrainment of miscellaneous media 4. Ingestion of plants or animals contaminated secondary to contact with contaminated miscellaneous media (e.g., exposed wastes and building materials)

exposed to contaminants through drinking water is more complex. First, the extent of water contamination must be determined. Is the contamination restricted to surface water bodies, aquifers, or both? How far have the contaminants traveled? The next step is to locate public and private water systems that use the contaminated water source. That information can be retrieved through the Federal Reporting Data System (FRDS). Retrievals can be requested for public water systems by state, county, or USGS hydrologic units. The FRDS database, maintained by the EPA Office of Drinking Water, includes the populations served and surface-water intake locations (8). In instances where groundwater is contaminated, private wells may also be a source of contaminated drinking water.

Because the FRDS does not contain data on private wells, an investigation of private well locations and the number of individuals served by them must be included when identifying populations exposed to contaminants via ingestion of groundwater. Other databases useful in identifying populations exposed through the ingestion of contaminated drinking water are also available. (Refer to section six of *Methods for Assessing Exposure to Chemical Substances. Volume 4: Methods for Enumerating and Characterizing Populations Exposed to Chemical Substances(8).)*

Populations exposed via ingestion of food. When uptake into plants and animals is possible, the health assessor should identify populations that are exposed or potentially exposed through consumption of contaminated plants and animals. The rate of consumption of contaminated plants and animals may differ considerably from the national average for certain populations living near hazardous waste sites. For example, families may consume homegrown vegetables as their main source of vegetables, or families may rely

on locally caught fish as a major source of protein. Other food sources that the assessor may consider include deer, rabbits, crayfish, cactus fruits, mushrooms, maple syrup, wild greens, and berries. When those foods are part of an exposure pathway at a site, the health assessor should identify the population at risk.

A survey or other adequate study of regional dietary habits may be necessary to determine the amount and frequency of contaminated food intake. Data on the number of licensed hunters and fishers for a particular geographic area may be found in surveys of state fish and game commissions. The number of persons exposed to contaminants after eating homegrown fruits or vegetables can be determined by estimating the percentage of households with fruit and vegetable gardens in the contaminated area (Appendix E).

After identifying exposed or potentially exposed populations, the assessor needs to determine site-specific factors that influence exposure frequency and duration.

6.5.2. Factors Influencing Exposure

Each site is unique and must be considered individually to determine the distinctive characteristics that could enhance or retard the frequency and magnitude of human exposure. In evaluating the likelihood of exposure, the assessor should consider the following factors:

Age of populations. Age of the population affects the type, level, and frequency of activities at or near the site. The presence of populations that may be at a higher risk, such as children or the elderly, should be noted.

Climatic conditions. A review of climatic conditions is necessary because it provides valuable information on the general types and frequency of outdoor and recreational activities of the local population. Subfreezing and other inclement weather, frozen ground, and frozen precipitation may serve as deterrents decreasing the frequency of human contact with soil contaminants.

Site accessibility. Humans can contact contaminated media if access to the site is not restricted or otherwise limited. The presence of a fence is not sufficient indication that the site is inaccessible. The assessor should determine the accessibility of the site, the contaminated materials (e.g., barrels) on or near the site, and the zones of contamination. Sites with abandoned buildings, standing water, or streams may attract children. Playgrounds and school yards near the site should be noted.

The assessor should also consider the extent of physical barriers, the condition of fencing, or security systems that may prevent (or inhibit) exposure. A review of local ordinances may reveal actions taken that minimize exposure, such as preventing the construction of private wells that could supply contaminated groundwater to local populations.

Land and resource use. A review of land use at or near the site will provide valuable information on the types and frequency of activities of the surrounding population and the probability for increased human exposure. Past, present, and future land use needs to be considered. Land use will significantly affect the types and frequency of human activities, thereby affecting the degree and intensity of contact with soils, water, air, exposed wastes, and consumable plants and animals. The assessor should pay particular attention to the following factors:

- **Residential areas.** Residential areas adjacent to the site provide potential for human exposure.

- **Recreational areas.** Attention should be given to recreational areas that may serve as points of exposure. Particular attention should be given to fields,

ditches, or physical hazards on or near the contaminated site that may attract children.

- **Food sources.** Agricultural and home fruit and vegetable production in contaminated soils present several routes of human contaminant exposure. The assessor should especially note when food production is used as a subsistence food source. Food crops or animal feed grown in contaminated soil may, through contaminant uptake, accumulation, and concentration by plants, be unsuitable for consumption by humans or by animals used for food production. In some areas, wild plants, animals, and fish may constitute a significant portion of the diet of local residents.

- **Surface-water use.** The assessor should verify the use of local surface-water bodies (public water supplies, recreation, irrigation and livestock feeding, and aquaculture). Information on water supply and proximity to water supply intakes that might be downstream from the site can be obtained from EPA or the state.

- **Groundwater use.** Similar information should be collected for groundwater use. Because more than one-half of the nation's population receives its drinking water from groundwater sources, it is particularly important that the assessor verify which public and private water supplies are affected by site contamination. To address this issue, the assessor may recommend a well-use survey, such as those performed by USGS or state or local health departments. Further, where it is believed that large municipal wells are drawing from contaminated aquifers for public water supplies, the assessor should seek information on location, depth, and construction of the wells. Pumping rates of high-capacity municipal, industrial, or agricultural wells can influence and to some extent determine localized groundwater flow regimes, often affecting contaminant transport within the capture zone of the well.

- **Private residential wells.** The assessor should not assume that because municipal water is supplied to a residential area no residents have private wells. The assessor should talk to local officials, such as the city or county planner, to determine the number and use of private wells that are or could be contaminated. If needed, the assessor should request that local or state officials conduct a well survey.

The assessor must use professional judgment in finalizing the list of exposure pathways. The exposure pathways of concern can be presented in the health assessment in a tabular format (such as Table 6.0.) indicating the contaminated media involved, potential receptors, points of exposure, and routes of exposure.

6.5.3. Estimation of Receptor Populations

Health assessors estimate the persons exposed or potentially exposed to contaminants for each exposure pathway. This estimate should be incorporated into the Pathways Analyses section of the health assessment for each completed and potential exposure pathway. Refer to Sections 6.6. and 6.7. and Example 15. in chapter 9.

Health assessors should use the following steps as a guide in estimating the receptor population:

1. Before the site visit, evaluate the completed and potential pathways (Sections 6.6. and 6.7.) from the available data.

2. During the site visit, further evaluate the exposure pathways by examining site access, use (work, play, riding, recreation, hunting, fishing), and local points of exposure.

3. After the completed and potential pathways have been identified, estimate the number of persons exposed or potentially exposed by each pathway. For example, if groundwater has been identified as a completed pathway, identify area groundwater use and determine the number of persons using municipal water or the number of persons using private wells that are contaminated or likely to be contaminated.

4. Consider contacting the following sources to determine the receptor population:

   - neighborhood associations;

   - individuals with municipal, county, and city agencies such as planners, managers, engineers, school officials, and health officials;

   - individuals at federal and state agencies such as parks departments, departments of natural resources, geologic surveys, and health agencies; and

   - surveys, census data, or other population information sources.

5. If an accurate number of persons cannot be obtained from the information sources in Step 4, estimate the numbers of persons by counting residences that represent a likely point of exposure in a completed or potential pathway. Multiply each residence by 2.5 persons (an estimate of 1990 census results). If a more accurate estimator is available, cite the source of the estimator and use that figure.

6. As a last resort, estimate the number of exposed or potentially exposed persons who live or work within a certain distance of the source of contamination or likely area of contamination off the site.

7. In the health assessment, describe the sources and method used to estimate the receptor population.

6.6. CATEGORIZING COMPLETED AND POTENTIAL EXPOSURE PATHWAYS

Exposure pathways can be categorized into completed exposure pathways or potential exposure pathways (Sections 6.6.1. and 6.6.2.) Each potential or completed exposure pathway represents a past, present, or future condition which should be disclosed by the health assessor. The assessor should discuss the likelihood of human exposure to site-related contaminants for each completed or potential pathway.

Although the major focus in the health assessment should be on the completed exposure pathways, site-specific conditions may suggest that special attention be given to one or more potential exposure pathways. When no completed pathways are present, the health assessor has the option of giving much more attention to potential exposure pathways. The health assessor must understand that each contaminant associated with a potential and completed exposed pathway will require further assessment in the Public Health Implications section.

6.6.1. Completed Exposure Pathways

A completed exposure pathway occurs when the five elements of an exposure pathway link the contaminant source to a receptor population. Should a completed exposure pathway exist in the past, present, or future, the population is considered exposed. This will be further explained in Section 6.6.3.

The health assessor should discuss each completed exposure pathway separately. Refer to Chapter 9 for an example of the format that should be used to describe completed exposure pathways in the health assessment. Present each completed exposure pathway by showing that each of the five elements (source of contamination, environmental media and transport, point of exposure, route of exposure, and receptor population) are connected.

Avoid confusing future completed exposure pathways with future potential exposure pathways by assessing the status of contamination at the point of exposure. Future potential exposure pathways are discussed in Section 6.6.2.

Assume that a future completed exposure pathway exists if each of the following apply:

1. Contamination currently exists (1) at a point of exposure or (2) in an environmental medium at a location that is likely to become a point of exposure within days, weeks, or months (e.g., undeveloped residential lots and vacant residential properties that have contaminated soil).

2. Persons in the community have continued, unrestricted access to a point of exposure (in the future), or may participate in activities that would expose them to contaminants in the environmental medium (e.g., constructing a residential playground on contaminated surface soil).

3. Institutional controls, building and zoning restrictions, and other ordinances do not exist to prevent contact with current contaminants at points of exposure or likely points of exposure where the probability for human contact with a contaminated medium is high, and can occur at any time in the near future.

For example, a future completed exposure pathway exists if a residence, residence under construction, or residential lot without a well lies above a contaminated aquifer and no restrictions exist that prevent the residents or property owners from drilling a well into the contaminated aquifer.

Another example of a future completed pathway is when present exposure is likely to continue in the future. If an occupied residence has contaminated yard soils, the residents would be the receptor population for a completed exposure pathway in both the present and the future. If a vacant residence or residence under construction has contaminated yard soils, a future completed pathway would exist because of the high probability of human contact in the near future. Designate an exposure pathway as a completed pathway in the future when a high probability exists for human activity/contact to occur in a contaminated medium at any time in the future. Keep in mind that present and future completed exposure pathways reflect present exposure, continued exposure, and exposure that is likely at any point in the future. Because completed pathways involve either actual exposure or a high probability for exposure in the future, give more attention to assessing and recommending measures to prevent current and future exposure in completed pathways. This approach will also assist the EPA in focusing on those pathways for which we have public health concerns. Determine how urgently ATSDR needs to have the information and data gaps filled to complete its assessment and how urgently public health actions and other follow-up activities need to be implemented. The Recommendations section should provide this sense of urgency.

Examples of human activities that could result in exposure from future completed pathways include the following:

- constructing wells and consuming contaminated groundwater;

- using contaminated soil areas for purposes (e.g., homes, schools, nurseries, and playgrounds) that would expose persons;

- consuming vegetables and fruits that have been grown in contaminated soil;

- consuming fish from contaminated waterways and water bodies;

- removing and/or using contaminated materials or media;

- occupying structures above areas with contaminated soil-gas; and

- residing or working in areas having contaminated indoor or outdoor air.

6.6.2. Potential Exposure Pathways

A potential exposure pathway exists when one or more of the five elements are missing, or if modeling is performed to replace real sampling data (e.g., modeled groundwater data using soil or other groundwater data levels). Potential pathways indicate that exposure to a contaminant could have occurred in the past, could be occurring now, or could occur in the future.

Assume a future potential exposure pathway exists if the contamination does not currently exist at a point of exposure. For example, the pathway is a future potential exposure pathway if the contamination (1) has to migrate to some point of exposure, or (2) has been modeled or projected to exist at some point of exposure. If a potable private well is projected to be affected by site-related contaminants (that exist upgradient of the well), a future potential exposure pathway exists.

In general, the discussion of potential exposure pathways should be brief. However, site-specific conditions, such as a highly contaminated groundwater plume upgradient of a public well used for a water supply, may warrant giving added attention to a potential pathway. The health assessment should include a list of the potential human exposure pathways with an estimate of the receptor population. Refer to Section 6.6.3. for instructions on presenting the summary information in a table format.

6.6.3. Summary Tables for Completed and Potential Exposure Pathways

The health assessors should summarize the exposure pathways, the five elements, and the pertinent time components for each pathway in one or more tables. Refer to Example 13, Completed Exposure Pathways, in Chapter 9 for a sample of how this information might be presented in the health assessment. Using the same or a similar table, the health assessor should make every effort to provide an estimate of the persons exposed and potentially exposed for each exposure pathway. Where possible, the table should include what contaminants affect each exposed and potentially exposed population. Refer to Example 15., Estimated Population For Completed and Potential Exposure Pathways, in Chapter 9 for a sample of how this information might be presented.

6.7. ELIMINATING AN EXPOSURE PATHWAY

Suspected or possible exposure pathways can be ruled out if site characteristics make past, current, and future exposure through the pathway extremely unlikely. Should site characterization show that some media are uncontaminated, the assessor can eliminate suspected exposure pathways associated with those media. However, the assessor should exercise caution in eliminating suspected exposure pathways based on uncontaminated environmental media. The quality and representativeness of the sampling and the likelihood for contamination of the media in the future need to be considered.

The assessor should not dismiss all exposure routes relating to a specific medium because one (or more) exposure route(s) pertaining to that medium are not of concern. For example, although inhalation of contaminants volatilized from contaminated soil may not be a problem at a site that is capped and has good vegetative cover, subsurface gas could still migrate and enter nearby homes. For those pathways that may be important to the public, the health assessor should explain why the contaminant of concern and the suspected pathway was eliminated. A suspected pathway can be eliminated if at least one element, which would link the five elements, is absent and will never be present. However, professional judgment

should be used when discussing pathways that may be eliminated if no environmental data exist.

If a suspected pathway cannot be categorized as a potential or completed exposure pathway, and no contaminants of concern have been identified, the pathway should be eliminated. Again, the health assessor should use professional judgment in deciding how to discuss this in the health assessment.

In general, the health assessor should avoid discussing environmental or human exposure pathways that are not supported by environmental data, site information, or the literature. If contaminants in suspected exposure pathways are considered important because of community health concerns, the health assessor should discuss them briefly in the health assessment.

6.8. DEFINING EXPOSURE

Three categories of exposure exist: exposed, potentially exposed, and no known exposure.

6.8.1. Exposed population

A population is considered exposed if a completed exposure pathway, which links a contaminant with a receptor population, exists in the past, present, or future. An exposed population includes persons who ingest, inhale, or contact site contaminants or are exposed to radiation in the past, present, or future. Examples of exposed persons include those who:

- have ingested, are ingesting, or will ingest the contaminant from one or more environmental media;

- have inhaled, are inhaling, or will inhale the contaminant from one or more environmental media;

- have contacted, are contacting, or will contact the contaminant in one or more environmental media; and

- were exposed, are exposed, or will be exposed to gamma radiation from one or more environmental media.

If an environmental medium (soil) contains a contaminant of concern at a point of exposure (a residential yard), and evidence already exists that a route of exposure (ingestion) has occurred, is occurring, or will occur, the health assessor should assume that persons living at that residence are exposed or will be exposed. If the residential yard contains a vacant house, the health assessor should assume that future residents will be exposed.

Persons should also be considered exposed if exposure has been verified by human biologic measurements or medical examination. For health assessments, human biologic measurements or medical examination are not necessary for the assignment of an exposure category to a population.

6.8.2. Potentially exposed population

A population is considered potentially exposed if a potential exposure pathway exists in the past, present, or future.

6.8.3. No known exposure

A health assessor assigns this category to a population if neither a completed exposure pathway nor a potential exposure pathway exists.

6.9. PRESENTING THE PATHWAYS ANALYSES SECTION

6.9.1. Preferred approach

In this approach, the Pathways Analyses are divided into the Completed Exposure Pathway subsection and the Potential Exposure Pathway subsection. The health assessor will provide a combined discussion of all five elements (source of contamination, environmental media, point of exposure, route of exposure, and receptor population) for each exposure pathway. This approach should make it easier for the health assessor to discuss each exposure pathway and how their elements are linked. In addition, this approach should allow the reading public to better understand how the five elements of an exposure pathway are linked. For additional discussion of the presentation of this approach, refer to Chapter 9.

6.9.2. Alternative approach

The alternative approach requires the health assessor to (1) include two subsections, Environmental Component and Human Exposure Component; (2) address the five elements of each exposure pathway (i.e., source of contamination, environmental media, point of exposure, route of exposure and receptor population); and (3) address how the five elements are linked for each exposure pathway. The first three elements represent the environmental component; the final two elements represent the human exposure component.

This two-subsection approach makes it more difficult to link all five elements for each exposure pathway. Because the two sections must address the environmental elements separate from the human exposure elements, it is difficult to understand the link between the five elements. For additional discussion of the presentation of this optional approach, refer to Chapter 9.

6.10. REFERENCES

1. In: Lyman WJ, Reehl WF, Rosenblatt DH, eds. *Handbook of chemical property estimation methods.* New York: McGraw-Hill Book Co., 1982.

2. Sax NI, Lewis RJ. *Hazardous chemicals desk reference*. New York: Van Nostrand Reinhold Company, 1987.

3. Thomas RG. Volatilization from water. In: Lyman WJ, Reehl WF, Rosenblatt DH, eds. *Handbook of chemical property estimation methods*. New York: McGraw-Hill Book Co., 1982.

4. Bingham FT, *et al.* Growth and cadmium accumulation of plants grown on a soil treated with a cadmium-enriched sludge. *J Environ Qual* 1975; 4(2):207-11.

5. Fries GF, Marrow GS. Chlorobiphenyl movement from soil to soybean plants. *J Agric Food Chem* 1981; 29:757-59.

6. Neely WB, Branson DR, Blau GE. Partition coefficient to measure bioconcentration potential of organic chemicals in fish. *Environmental Science and Technology* 1974; 8:1113-5.

7. Varanasi U, Gmur DJ. Hydrocarbons and metabolites in English sole exposed simultaneously to benzo[a]pyrene and naphthalene in oil-contaminated sediment. *Aquatic Toxicology* 1981; 1:49-67.

8. EPA Office of Toxic Substances. *Methods for assessing exposure to chemical substances. Vol. 4: Methods for enumerating and characterizing populations exposed to chemical substances.* Washington, D.C.: Environmental Protection Agency, 1986. EPA 560/5-85-004 and PB86-107042.

9. Sedman RM. The development of applied action levels for soil contact: a scenario for the exposure of humans to soil in a residential setting. *Environmental Health Perspectives* 1989; 79:291-313.

10. EPA Office of Emergency and Remedial Response, Office of Solid Waste and Emergency Response. *Superfund exposure assessment manual.* Washington, DC: Environmental Protection Agency, 1988; EPA 540/1-88/001.

11. EPA Office of Emergency and Remedial Response. *Risk assessment guidance for superfund. Volume 1. Human health evaluation (Part A).* Washington, DC: Environmental Protection Agency, 1989; EPA 540/1-89/002.

7. DETERMINING
PUBLIC HEALTH IMPLICATIONS

The preceding sections deal with steps in the health assessment process that help the assessor: 1) identify site characteristics and conditions that lead to contaminant release and transport; 2) identify contaminants of concern at the site; and 3) evaluate the site's human exposure potential. Health assessors must now link the site's human exposure potential with health effects that may occur under these site-specific conditions or that may have occurred in the past. This step in the health assessment process is presented under **Public Health Implications**. This section consists of three major subsections:

Toxicologic Evaluation

Health Outcome Data Evaluation

Community Health Concerns Evaluation

7.1. TOXICOLOGIC EVALUATION

In this section of the health assessment, the assessor uses the previously gathered information on contaminants of concern and exposure pathways to estimate potential exposures to site contaminants. The assessor then determines the potential health effects that could result from these exposures. The assessor must consider numerous medical, toxicologic, demographic, and environmental factors that might affect the impact, on human health, of exposures to hazardous substances. The following elements should be included in this analysis:

- estimating exposures;

- comparing exposure estimates with health guidelines;

- determining exposure-related health effects;

- evaluating factors that influence adverse health outcomes; and

- determining health implications of physical and other hazards (e.g., hazards related to fires, drowning, falling).

7.1.1. Estimating Exposures

An estimate of potential exposures to hazardous substances should be derived using site-specific exposure information. In the absence of site-specific information, the health assessor should refer to standard exposure estimates, such as those contained in EPA's *Exposure Factors Handbook* (1).

The estimation of exposures should focus on current conditions at the site. However, if possible, past exposures to hazardous substances should also be considered because they may affect the current health status of previously exposed individuals. For example, past exposures to lead may have resulted in the storage of lead in bones, which could be released during pregnancy, menopause, or other metabolic stresses (2). Therefore, previously exposed individuals may be at risk for lead toxicity even though they are not currently exposed.

The health assessor should also consider the potential for future exposures from the site, the impact of site removal and remediation activities, or a change in use of the site. When estimating exposure at a site, the following factors should be considered:

Exposure Duration. To determine the overall and incremental risk posed to an exposed population, the assessor must try to determine how long the population has been exposed to site contaminants.

By examining the site's history, the assessor usually can adequately define the initial and final dates of operation or receipt of wastes. With these dates in mind, an estimate of the maximum duration of exposure may be possible. However, this time frame cannot account for exposure that continued after the site was closed (e.g., via contaminated surface- or groundwater).

In addition, dates of emergency removal or remedial action, closure of water supplies, erection of physical barriers or security systems to prevent access, and public notification of site contamination may also be used to estimate when public exposure is likely to have ceased. When contaminated water supplies are under investigation, information on installation of public or private water supplies, along with the construction dates of residences and neighborhoods, can also help to estimate the length of exposure.

Exposure Frequency. Exposure frequency is the amount of time an individual has access to a contaminated area and reflects the time for possible exposure. Exposure frequency can be estimated as the ratio of the average number of hours an individual is exposed per day to the number of hours in a day, or the number of days an individual is exposed per week to the number of days in a week.

Exposure Fluctuation. The health assessor should consider whether the exposure is continuous or intermittent. The same total dose of a chemical can cause different toxic effects depending on whether the dose is administered over a short or prolonged period. The historical development of a site and changes in the environment over time also influence exposure.

Bioavailability. In order to exert a toxic effect, most chemicals must be absorbed into the body. For some chemicals, there may be quantitative data that allow for a comparison of the bioavailability of the chemical in experimental animals and humans. However, for most chemicals, it is assumed that bioavailability is the same in animals and humans.

Absorption of contaminants may vary dramatically depending on the route of exposure. Contaminants that are readily absorbed through the gastrointestinal tract may not be readily absorbed through the respiratory tract and vice versa. Therefore, one cannot assume that absorption will be equivalent by different routes of exposure. In general, ATSDR does **not** recommend using experimental data obtained from one route of exposure to calculate doses of toxic effects for a different route of exposure.

Once a contaminant has been absorbed into the body, it is distributed to various organs in accordance with pharmacokinetic principles. The dose that reaches the target organ is what ultimately determines the toxic effect. Physiologically based pharmacokinetic models have been developed to estimate dose levels in various body compartments and organs; however, such models are not routinely used for health assessment purposes.

Appendix D contains detailed information for estimating exposures from various pathways. When estimating exposures, the health assessor should specify whether the estimates are based on maximum contaminant concentrations, an average of measurements taken from the same location, or the range of contaminant concentrations detected. The health assessor should recognize that use of the maximum detected concentration of a contaminant to calculate the exposure dose may result in an overestimate of actual exposure.

Estimates of exposure dose are generally determined for exposure to a single contaminant via a single route of exposure. However, at many sites, exposure to a contaminant may occur through multiple routes of exposure. When this occurs, the exposures from the various pathways should be summed to derive a total exposure dose.

7.1.2. Comparing Exposure Estimates With Health Guidelines

After estimating exposures at the site, the assessor must determine whether these exposures are of concern. The first step in identifying exposures of concern is to compare the exposure of interest with guidelines that are designed to protect human health. Appendix B discusses regulatory standards and guidelines that may be useful in such operations. As a general rule, if the guideline is exceeded, the exposure is of potential concern. However, sometimes additional medical and toxicologic information may indicate that exposures exceeding these guidelines are not of health concern. In other instances, exposures below guideline values could be of health concern because of interactive effects with other chemicals or because of the increased sensitivity of certain individuals. Thus, additional analysis (see Section 7.4.) is necessary to determine whether health effects are likely to occur. Nonetheless, use of environmental guidelines and medical or toxicologic health guidelines is the first step in making this determination.

Environmental Guidelines

The term "environmental guideline" refers to health-related guidelines that represent acceptable medium-specific concentrations (e.g., mg/L) of a compound to which humans may be exposed via a specific exposure pathway (e.g., drinking water). Environmental guidelines include occupational—regulatory and nonregulatory—guidelines that have been established by EPA and other organizations (e.g., American Conference of Governmental Industrial Hygienists [ACGIH], Occupational Safety and Health Administration [OSHA], and National Institute for Occupational Safety and Health [NIOSH]). These guidelines, based on specific, usually well-known exposure

scenarios and target populations, are convenient to use because they are familiar to the scientific and regulatory communities. However, it should be noted that some of these guidelines may not be supported by an adequate toxicity base. Furthermore, these guidelines may not be appropriate to use for sensitive individuals (the young, the old, the ill) or for continuous exposures.

When environmental guidelines are available for the site's contaminants of concern, these guidelines should be compared with site concentrations. When the concentration of a contaminant exceeds an environmental guideline, the concentration and guideline should be reported in the health assessment. In addition to environmental guidelines, the assessor should use medical and toxicologic health guidelines to assess exposures.

Medical or Toxicologic Health Guidelines

To evaluate whether the contaminants of concern are likely to pose a health threat under site-specific exposure conditions, the assessor should compare estimates of exposure dose with health-based values such as ATSDR's Minimal Risk Levels (MRLs) or other reference values described in Appendix B. An MRL is defined as an estimate of the daily human exposure to a substance that is likely to be without an appreciable risk of adverse, non-cancer health effects over a specified duration of exposure.

These health guideline values provide perspective on the relative significance of human exposure to contaminants at the site. These values alone, however, cannot be the sole determination of the potential health threat of a particular chemical.

When comparing environmental contaminant concentrations to regulatory standards, the assessor must consider the assumptions used to derive the values to determine if they are applicable under site-specific conditions. The assessor should also consider the potential for cumulative contaminant doses from multiple routes of exposure.

When simultaneous exposure to multiple chemicals occurs, there is a potential for additive, synergistic, or antagonistic interactive effects. If it is concluded that interactive toxic effects may result in a public health hazard, the health assessor should provide data and evidence to support that conclusion.

The assessor should also consider information suggesting that the health impact of a contaminant may be modified by host-specific factors such as nutritional deficiencies, life-style factors, age, sex, or preexisting disease. Information on toxicological interactions and host-specific modifying factors is contained in the ATSDR Toxicological Profiles.

Health guideline values are usually derived from experimental animal data, based on broad assumptions, and corrected by a series of uncertainty factors. Thus, the values serve only as guidelines and not as absolute values that explicitly divide ranges of safety from ranges of risk.

The assessor's reliance on such values should also be tempered by an awareness of toxicologic information used in setting guidelines for particular contaminants. Studies of the adverse effects of exposure to hazardous substances may be restricted to a limited number of end points.

Many experimental animal studies focus on carcinogenic effects. Other studies may be designed to examine a single toxic end point (e.g., developmental toxicity). Regulations for acute or occupational exposures may be based on a specific type of adverse response (e.g., eye or skin irritation) and may not be protective of other adverse health effects (e.g., central nervous system depression or neurobehavioral effects). Therefore, the assessor needs to evaluate all available data for a specific chemical.

If guidelines are not available for the specific exposure route of concern at a site, criteria developed for other exposure routes may be used. However, care should be exercised when drawing conclusions from those comparisons. For example, when standards are not available for dermal contact or for inhalation, reference doses (RfDs) or cancer slope factors derived for ingestion exposures may be used, but consideration should be given to the validity of such extrapolations.

7.1.3. Determining Exposure-Related Health Effects

It is important for the health assessor to use the best medical and toxicologic information available to determine the health effects that may result from exposure to contaminants at a site. Such information can be derived from ATSDR's chemical-specific Toxicological Profiles, standard toxicology textbooks, and scientific journals of environmental toxicology or environmental health. Assessors should also consult on-line databases, such as the Hazardous Substances Data Bank (HSDB) and Toxline, for the most current toxicologic and medical information.

The assessor should indicate in the health assessment whether health concerns are for acute, intermediate, or chronic exposures. In the ATSDR Toxicological Profiles, acute exposures refer to those of 14 days or less; intermediate exposures are from 15-364 days; and chronic exposures are for 365 days or more.

When evaluating the health impact of exposure to hazardous substances, the assessor should consider data from studies of human exposures as well as from the results of experimental animal studies. For health assessment purposes, the use of human data is preferred because it eliminates uncertainties involved in extrapolating across species. However, human data are often unavailable, particularly for chronic, low-dose exposures. Furthermore, adequate human data are often not available to establish a dose-response relationship. In the absence of adequate human data, the health assessor must rely on the results of experimental animal studies.

When evaluating the potential health effects of exposures to hazardous substances, the health assessor should focus on health effects relevant to the contaminant doses that could result from exposures at the site. In addition to considering the contaminant dose, the health assessor should consider the frequency of exposure and the exposure duration. At many hazardous waste sites, exposures can often be characterized as chronic and of a low dose. Health effects data and information for such exposures are often lacking. In those instances, the health assessor may have to rely on studies that involve shorter exposures and/or higher dose levels. If such studies are used as the basis for a health assessment, the assessor should acknowledge the qualitative and quantitative uncertainties involved in those extrapolations.

ATSDR recommends that site-related exposures be compared to studies or experiments involving comparable routes of exposure—i.e., ingestion, inhalation, and dermal contact. However, in some instances, it may be necessary to use data from studies based on different exposure pathways. Caution should be used when drawing conclusions from such studies because of uncertainties involved in route-to-route extrapolations due to differences in contaminant absorption, distribution, metabolism, and excretion. In addition, a contaminant might exert a toxic effect by one route of exposure, but not by another (e.g., chromium is reported to be carcinogenic by inhalation but not by ingestion).

Information on the toxic effects of chemical exposure in humans and experimental animals is contained in the ATSDR Toxicological Profiles. These documents also contain dose-response information for different routes of exposure. When information is available, the Toxicological Profiles also contain a discussion of toxic interactive effects with other chemicals and a description of potentially sensitive human populations.

7.1.4. Evaluating Factors That Influence Adverse Health Outcome

The assessor should review factors that may enhance or mitigate health effects resulting from exposure to site contaminants. The assessor should consider other medical and toxicologic information, health implications for sensitive subpopulations, health implications of past and future exposures, and the effects of site remediation on human exposure.

Health Implications of Other Medical and Toxicologic Factors

The health effects identified by comparing dose estimates with guideline values and referring to health effects information (in Sections 7.3. and 7.4.) should also be evaluated on the basis of other toxicologic and medical factors that can enhance or mitigate effects of exposure. When relevant, these factors should be identified and their health implications discussed in the health assessment. Factors the health assessor may consider are these:

- distribution within the body (the fate of the chemical after ingestion, inhalation, or dermal contact);

- target organs (site of major toxicity);

- toxicokinetics of substance (including possible transfer to cow's milk or nursing mother's milk);

- enzyme induction (chemical induction of various enzyme systems may increase or decrease chemical toxicity);

- the cumulative effect of exposures to chemicals that bioaccumulate in the body (e.g., lead, cadmium, organochlorine pesticides);

- chemical tolerance (decreased responsiveness to a toxic chemical effect resulting from previous exposure to that chemical or to a structurally related chemical);

- immediate versus delayed effects (effects observed rapidly after a single exposure versus effects that occur after some lapse of time);

- reversible versus irreversible effects (ability of affected organs to regenerate);

- local versus systemic effects (whether the effect occurs at the site of first contact, or must the chemical be absorbed and distributed before the effect is observed);

- idiosyncratic reactions (genetically determined abnormal reactivity to a chemical that is qualitatively similar to reactions found in all persons, but may take the form of either extreme sensitivity to low doses or extreme insensitivity to high doses);

- allergic reactions (adverse reaction to a chemical resulting from previous sensitization to that chemical or a structurally related one); and

- other related disease effects (effect of chemical on previously diseased organ).

In addition to the medical and toxicologic factors identified here, the health assessor should consider population-specific factors that may enhance or mitigate health effects associated with exposure to the contaminants of concern.

Health Implications for Subpopulations

Many subpopulations may be identified at a site. Each subpopulation has special concerns that must be considered when determining public health implications.

Sensitive Subpopulations

Perhaps the most crucial set of factors that the assessor must weigh are those that influence differential susceptibility to the effects of specific compounds. Age, sex, genetic background, nutritional status, health status, and general lifestyle may each influence the effects of contaminant exposure. The assessor should carefully consider the impact that each of these factors may have at a specific site for a given population.

Age and Sex. Age-related susceptibility to toxic effects is more widespread than many public health workers realize. Although other factors generally affect only a small segment of the population at a given time (e.g., only 4% of the total population are carriers of hereditary DNA-repair diseases), everyone is at enhanced risk because of age factors at some point in his or her lifetime (e.g., all infants are at enhanced risk for radiation-induced cancer).

EPA and other federal agencies have acknowledged that the very young are a particularly high-risk group that must be protected more stringently from the adverse effects of certain compounds. For example, the EPA primary drinking water standard for nitrate was set to protect the most susceptible high-risk group: infants in danger of developing methemoglobinemia. Similar age-related sensitivities are reflected in levels set for lead in ambient air and drinking water, and for mercury in aquatic systems.

The very young are not always the age group of most enhanced risk. In some instances, adults are at greater risk of toxicity than infants or children (3). For example, the young seem more resistant than adults to the adverse effects of renal toxicants such as fluoride and uranyl nitrate. The recent acknowledgement that elderly subpopulations may have significantly heightened susceptibility to contaminants because of lower functional capacities of various organ systems, reduced capacity to metabolize foreign compounds, and diminished detoxification mechanisms is also an important consideration.

Finally, some adverse health effects may be mediated by hormonal influences and other factors that are sex-linked. In general, sex-linked differences in toxic susceptibilities have not been extensively investigated. However, it is well documented that pregnant women are often at significantly greater risk from exposure to beryllium, cadmium, lead, manganese, and organophosphate insecticides than other members of the general population because of various physiologic modifications of pregnancy (4). A developing fetus is at greater risk from compounds that exert developmental effects.

Biochemical or Genetic Susceptibilities. The presence of subpopulations with certain inherent biochemical or genetic susceptibilities should also be considered when evaluating the potential health threats of a site. Studies by the National Academy of Sciences (5), Stanbury et al (6), and Stokinger and Scheel (7) indicate that genetic predisposition is a determining factor in as many as 150 diseases. Studies of some of these genetically determined diseases have shown an increased susceptibility to the toxic effects of certain pollutants. For example, certain percentages of various ethnic groups are known to suffer from inherited serum alpha-1-antitrypsin deficiency (3,8), which predisposes them to alveolar destruction and pulmonary emphysema (9,10,11). Persons with this deficiency are especially sensitive to the effects of certain pollutants. Such information can be used in conjunction with information on the ethnic makeup of surrounding populations to better evaluate potential toxic effects associated with a site.

In addition, persons who have chronic diseases may also be at increased risk from exposure to certain contaminants. Individuals with cystic fibrosis are less tolerant of the respiratory and gastrointestinal challenges of some pollutants. Persons with hereditary blood disorders, such as sickle-cell anemia, have increased sensitivity to compounds such as benzene, cadmium, and lead, which are suspected "anemia producers" (4,7). Those examples should alert the health assessor to the importance of determining (from documents or during the site visit) the presence and proximity of hospitals or convalescent homes where sensitive subpopulations are likely to be found.

Socioeconomic Factors. Demographic and land-use information should also help identify the relative socioeconomic status of exposed populations. This information may provide important clues for this step of the health assessment process. Not only can socioeconomic status be an important indication of susceptibilities to specific pollutants, but such information may also help identify confounding nutritional deficiencies or behaviors that enhance sensitivity to toxic effects. Studies have shown that dietary deficiencies of vitamins A, C, and E may increase susceptibility to the toxic effects of polychlorinated biphenyls, chlorinated hydrocarbons, some pesticides, ozone, and other substances. Other studies have indicated that deficiencies in trace metals such as iron, magnesium, and zinc exacerbate the toxic potential of fluorides, manganese, and cadmium.

Populations with sensitivities due to nutritional deficiencies may be suspected in areas of low socioeconomic status and extreme poverty or in areas with large numbers of indigents. Elderly populations have also been identified as a subgroup at risk of susceptibility because of nutritional deficits.

The assessor must carefully examine demographic information for particular groups on or near the site who might be especially sensitive to toxic effects. Any suspected high-risk groups should be specifically identified in the health assessment report. The health assessor should also note that information on the number and proximity of people in high-risk subpopulations is vital for providing optimal public health protection. Locations of schools, playgrounds, recreational areas, and retirement or convalescent homes on or near a site should be carefully noted as important indications of the presence of sensitive subpopulations. Enumeration of ethnic groups within the population and characterization of socioeconomic status may also indicate sensitive subpopulations near a site. When those groups are known to be at risk from exposure to site contamination, the assessor should determine from available medical and toxicologic literature and databases the nature and magnitude of adverse health effects likely to result.

Worker and Residential Subpopulations

Some sites may have worker populations on-site. The health assessment should clearly identify health concerns for both off-site residential and on-site worker populations. The health assessment should also address health implications for workers involved in site remediation. Furthermore, health assessors should consider the families of workers who may be (or have been) exposed through contact with work clothing or other secondary means.

Site-Specific Exposed Subpopulations

During the assessment process, subpopulations of special concern should be identified. Those individuals may be at increased risk because of greater sensitivity, compromised health status, concomitant occupational exposures, or other reasons. If such individuals exist, they should be identified in the health assessment, and appropriate recommendations should be made for their protection.

Health Implications of Past Exposures

When determining the health implications of a site and addressing the community's health concerns, the assessor should consider past, current, and potential future exposures.

Past exposures are difficult to address because they are difficult to quantify; nonetheless, significant exposure may indeed have occurred. When addressing community health concerns about past exposures, the assessor should review community-specific health outcome databases, such as morbidity data and disease registries, to evaluate a correlation between past and current health outcomes and past exposures. When past exposures have been documented and health studies have not been performed, the assessor should consider recommending the site for health effects studies or performing a review of community health records.

Health Implications of Future Exposures

An important aspect of public health implications at a site involves differentiating between current and future completed exposures, and current and future potential exposures. When considering completed and future potential exposures, the assessor should decide whether other exposures are possible from continued contaminant migration, anticipated (or likely) land development, or remedial activities. The assessor should make appropriate recommendations to mitigate future exposures.

Health Implications of Site Remediation

When determining the health implications of a site, the assessor should consider the effect(s) of remedial activities. Previous, current, or planned remedial activities can significantly affect conclusions about site-related health concerns. When removals or emergency response measures have occurred previously, the assessor should consider what effect those measures have had on health. Similarly, if site remediation is already occurring or has been announced, the assessor should determine what likely effect it will have. The health assessment should be responsive to community health concerns about remedial activities. In addition, discussions in the health assessment about exposure pathways should clearly identify and differentiate between pathways that are currently present and pathways that may have occurred in the past—but have been eliminated or significantly reduced by remedial activities.

7.1.5. Determining Health Implications of Physical Hazards

Besides the health hazards posed by the site's chemical contamination, physical hazards may also have been identified during the site visit. The assessor should make appropriate recommendations and conclusions in the health assessment depending on the significance of the physical hazards with respect to the site. For example, an abandoned site may be subject to uncontrolled dumping of household or construction debris, which may be considered a physical hazard, depending on the location of the site, type of debris, and surrounding land use.

7.2. EVALUATING HEALTH OUTCOME DATA

The health assessor should evaluate available health outcome data for all identified plausible outcomes and outcomes of community concern as appropriate. First, the health assessor should identify appropriate health outcomes for evaluation by using the previously developed environmental and toxicological information and applying the criteria outlined in Section 7.2.2. Further guidance in evaluating and discussing health outcome data in health assessments is presented in Section 7.2.4.

This section discusses the following issues:

- use of health outcome data in the health assessment process;

- criteria for evaluating health outcome data;

- using health outcome data to address community health concerns; and

- guidance for evaluating and discussing health outcome data in health assessments.

7.2.1. Uses of health outcome data in the health assessment process

Health outcome data can be used for many purposes. The data can provide valuable information on patterns of specified outcomes such as infant mortality and specific types of cancers. Those data can also be used to compare the prevalence of a specific outcome among different populations (e.g., town A compared with town B, or with the state or the country).

Health outcome data can provide information on the general health status of the community living near a hazardous waste site. Evaluating health outcome data also assists in addressing community health concerns such as: "Are more members of the community living near a specified site suffering from a specified disease compared with another population not living near this site." The evaluation of these data is not meant to and cannot establish cause and effect (i.e., without consideration of other important factors, such as exposure, biologic plausibility, and other causes, an elevated brain cancer rate alone cannot be considered conclusive evidence that living near a waste site is the sole cause for a specific health outcome in the nearby community).

In addition to addressing community health concerns, health outcome data can provide guidance in determining appropriate follow-up health actions.

7.2.2. Guidance Criteria for evaluating health outcome data in the health assessment process

General Guidelines

Health outcome data should be evaluated for plausible carcinogenic and non-carcinogenic outcomes, based on the nature and extent of exposures and the adverse toxicologic and physiologic health outcomes resulting from those exposures.

Alternatively, an evaluation of health outcome data may be undertaken to assist health assessors in addressing community health concerns. The outcome(s) of concern to the community may or may not be plausible and/or related to the exposures associated with a site.

Plausible Outcomes

Plausible Carcinogenic Outcomes

The following is a stepwise approach to evaluating health outcome data related to carcinogenic endpoints.

Step 1:

Establish the existence of a completed exposure pathway.

Each completed exposure pathway exists of five elements: (a) source of contamination, (b)

environmental transport mechanism (environmental medium), (c) point of human exposure, (d) route of exposure, and (e) the presence of a receptor population.

Step 2:

Determine whether chemicals associated with the completed exposure pathway(s) are carcinogens.

- Identify Carcinogens Appropriate for Data Analysis.

Each chemical in a completed exposure pathway should be evaluated based on **sufficient** human evidence, **limited** human evidence, or **sufficient** animal evidence of carcinogenicity. Designating a chemical as a carcinogen (for purposes of health outcome data evaluation) is based on these determinations:

- classification by the National Toxicology Program (NTP)[1] in its Annual Report on Carcinogens as a **known human carcinogen** or **reasonably anticipated to be a carcinogen**; or

- classification by the International Agency for Research on Cancer (IARC)[2] as a 1, 2A, or 2B carcinogen; or

- classification by the Environmental Protection Agency (EPA)[3] as an A, B1, or B2 carcinogen.

---

1   The National Toxicology Program in its Annual Report on Carcinogens classifies a chemical as a "known human carcinogen" based on sufficient human data. Its classification of a chemical as being "reasonably anticipated to be a carcinogen" is based on limited human or sufficient animal data.

2   IARC defines a class 1 carcinogen as a substance for which studies in humans indicate a causal relationship between the agent and human cancer. Class 2 carcinogens are those reasonably anticipated to be carcinogens. For a 2A classification, there is limited evidence of carcinogenicity from human studies that indicates a causal interpretation is credible, but not conclusive. A 2B classification indicates that there is sufficient evidence of carcinogenicity from studies in experimental animals.

3   In EPA's classification scheme, a chemical is considered a class A or human carcinogen based on sufficient evidence from studies of humans. A substance is considered class B1 if there is limited evidence from human studies. B2 is used when evidence for carcinogenicity is inadequate or non-existent based on human studies, but sufficient based on animal studies.

Information on the carcinogen classification for a specific chemical can be found in Sections 2.2. and 2.4. of the ATSDR Toxicological Profile for that chemical. Additional sources of information are the Annual Report on Carcinogens published by the National Toxicology Program and the IRIS and HSDB databases in the National Library of Medicine's TOXNET database.

- For each completed exposure pathway involving a carcinogen, estimate when the exposure began. A latency period of at least 10 years between exposure and diagnosis has been observed in most studies of human cancer.

Step 3:

Evaluate cancer outcomes for all organ sites.

Cancer outcomes should be obtained for all organ sites based on carcinogens selected in step 2. IARC has recommended that *in the absence of adequate data on humans, it is biologically prudent and plausible to regard agents for which there is sufficient evidence of carcinogenicity in experimental animals as if they presented a carcinogenic risk to humans* [IARC 1987]. However, animal evidence does not consistently establish the cancer site in humans, so it cannot be used for this purpose. Because of those uncertainties, it is prudent public health practice to obtain cancer data for all organ sites.

Step 4:

Estimate the number of individuals being exposed.

For each chemical carcinogen and completed exposure pathway, estimate the number of exposed individuals and note where they live; this information is essential for obtaining appropriate cancer outcome data.

Step 5:

Identify available health outcome databases.

Identify and characterize databases which include the exposed population. For example, note the following characteristics:

- type of database (e.g., vital statistics, cancer registry, completed studies, other health records known to be available);

- name of data source (e.g., West Alabama Regional Cancer Registry);

- responsible organization (e.g., University of Alabama);

- lowest geographic unit of analysis (e.g., region, state, county, city/town, zip code, census tract, block);

- years analyzed (e.g., 1980-1985); and

- demographics (e.g., age, sex, race, occupation, education, income).

Appendix H provides a checklist to assist health assessors in characterizing health outcome databases listed in Appendices F and G. Appendix I provides information on the major characteristics of frequently used secondary health databases.

Plausible Non-Carcinogenic Outcomes

The health assessor should evaluate health outcome data for plausible non-carcinogenic endpoints using the following steps:

Step 1:

As with the carcinogenic outcomes, the health assessor should establish whether a completed pathway exists.

Step 2:

Identify plausible non-cancer outcomes based on the toxicologic characteristics of the chemical(s) in the completed exposure pathway.

Information on chemical-specific non-carcinogenic outcomes can be found in Sections 2.2. and 2.4. of ATSDR's Toxicological Profiles.

Step 3:

Select appropriate outcomes to be evaluated.

Select all health outcomes identified in Step 2 as plausible adverse health effects in humans. In addition, when possible, the health assessor should evaluate health outcomes identified in animals that can be reasonably expected to occur in humans. Such instances include scenarios where the outcome of interest was observed in repeated studies in multiple types of animals, and the exposure route in these animal studies is the same as the human exposure route being investigated.

Step 4:

Estimate the number of individuals being exposed.

Repeat Step 4 under Carcinogenic Outcomes.

Step 5:

Identify available health outcome databases.

Repeat Step 5 under Carcinogenic Outcomes.

7.2.3. Using Health Outcome Data to Address Community Health Concerns

Community health concerns constitute one of the three main data sources on which health assessments are based. During the health assessment process, community concerns are identified by the health assessor. These concerns are gathered from a variety of sources: individual citizens, activist groups, politicians, local and state health departments, and local media.

Evaluating health outcome data may assist in addressing health-related community concerns. Specifically, these data are helpful in addressing concerns related to a high incidence of a specific disease in a community living near a hazardous waste site.

To address such concerns, the health assessor should follow these steps:

Step 1:

Determine whether the outcome of concern is plausible. If the outcome is plausible, the health assessor should follow the procedures outlined in Section 7.2.2.

Step 2:

If the outcome is not plausible, the health assessor should initiate a search for relevant health outcome databases.

When addressing this health concern, the health assessor should clearly state that the evaluation of health outcome data was to address a specific community health concern, and that the outcome under investigation is unlikely to be associated with exposure to site-related contaminants.

Step 3:

If no health outcome data are available to address a specific outcome of concern to the community that is not likely to be biologically plausible, the health assessor should address it using the pathway analysis information and toxicologic data.

7.2.4. Guidance in evaluating and discussing health outcome data in health assessments.

The health assessment format currently contains two health outcome data components: Health Outcome Data and Health Outcome Data Evaluation.

Health Outcome Data

Health Outcome Data is presented in Subsection D of the background section of the health assessment. The purpose of this subsection is to list all health outcome databases and information evaluated in the health assessment.

This list can only be compiled after the health assessor has identified the appropriate databases to be used by applying the criteria outlined in this chapter. The list should include the names of the databases and the sources and/or organizations maintaining them. It is not appropriate to discuss the evaluation and findings of health outcome data in this subsection.

Health Outcome Data Evaluation

Health Outcome Data Evaluation is presented in Subsection B in the Public Health Implications section of the health assessment. The purpose of this subsection is to discuss and present the findings of the evaluation of all biologically plausible health outcome data -- carcinogenic and noncarcinogenic -- and outcomes of concern to the community regardless of plausibility. Again, to identify the plausible outcomes, the health assessor should use the criteria outlined in this chapter.

As discussed previously, health outcome data used in health assessments cannot, in most cases, establish a causal link between exposure and adverse health effects in individuals living near a hazardous waste site.

However, health outcome data can:

- compare the occurrence of a specific health outcome (disease) among populations; and

- assist the health assessor in addressing community concerns related to increased occurrence of disease in the community.

The evaluation of any health outcome database should include:

a. a discussion of the characteristics of the database under evaluation (e.g., number of years for which data are available, smallest geographic unit for which data are collected, which years of data were included in the analysis, etc.);

b. a discussion on how the population associated with the site relates to the smallest geographic unit for which data are available (e.g., the community is part of a city, but data are only available at the county level);

c. if comparison populations are used (e.g., the city, county, state), a discussion of the characteristics of the populations;

d. a discussion on which methods of analysis will be used (crude rates, standard morbidity/mortality ratios, adjusted rates, etc.);

e. a discussion of the findings of analysis;

f. a discussion of the limitations of the findings. This should include discussing the database-specific limitations as well as those limitations associated with the analysis methods used. Such limitations include:

- the reliability of the data collected as it relates to the duration of existence of the database (in general, the older the database, the more reliable the data);

- the role of the smallest geographic unit for which data are collected in assessing the presence of increased rates of a specific health outcome for a community that is smaller than that residing in the smallest geographic unit; and

- the implications of using crude rates versus adjusted rates.

g. a discussion of how the results of the analysis relate to the overall public health implications of the hazardous waste site under evaluation; i.e., what do elevated and (non)elevated rates of a health outcome—plausible, non-plausible, or of community concern—mean with respect to the overall impact of the site on the nearby community?

As with other types of information used in the health assessment, the health assessor should consult with appropriate staff within his/her organization when evaluating health outcome data.

7.3. EVALUATING COMMUNITY HEALTH CONCERNS

The evaluation of community health concerns is presented as Subsection C of the Public Health Implications section of the health assessment. In this subsection, the health assessor should address every health concern expressed by the community as listed in the Community Health Concerns section of the health assessment. To address each outcome of concern, the health assessor should use environmental contamination data, exposure pathways analyses, and health outcome data as appropriate.

First, the health assessor should determine if the outcome(s) of concern is (are) biologically plausible. Because all plausible outcomes have already been evaluated, the health assessor can simply refer to that discussion. If the outcome(s) of concern is (are) not plausible, the health assessor may initiate the identification and evaluation of health outcome data, clearly indicating that this evaluation is being undertaken to help address a specific concern. If no health outcome data are available to address a non-plausible concern, the health assessor should discuss relevant pathway analysis and toxicologic and medical information to address this concern. It may also be helpful to follow this discussion with a brief general description of the outcome (disease) of concern. The description can include where in the human body the disease occurs, the prevalence of the disease, the (multiple) cause(s) of the disease, etc.

When addressing community health concerns in health assessments, it is helpful to restate the concern as expressed by the community, rather than as interpreted by the health assessor.

Having identified the site's health implications, the assessor is ready to complete the conclusions and recommendations that have been developed during the health assessment process.

7.4. REFERENCES

1. EPA Office of Health and Environmental Assessment. *Exposure factors handbook.* Washington, DC: Environmental Protection Agency, March 1990; EPA/600/8-89/043.

2. ATSDR Division of Toxicology. *Toxicological Profile for Lead.* Atlanta, GA: Agency for Toxic Substances and Disease Registry, June 1990; TP-88/17.

3. Calabrese EJ. *Age and susceptibility to toxic substances.* New York: J.H. Wiley and Sons, 1986.

4. Calabrese EJ. *Pollutants and high-risk groups: the biological basis of increased human susceptibility to environmental and occupational pollutants.* New York: J. H. Wiley and Sons, 1978.

5. National Academy of Sciences (NAS). Special risk due to inborn error of metabolism. In: *Principles for evaluating chemicals in the environment.* Washington, DC: NAS Press, 1975:331.

6. Stanbury JB, Wyngaarden JB, Fredrickson DS, eds. *The metabolic basis of inherited disease.* New York: McGraw-Hill Book Company, 1972.

7. Stokinger HE, Scheel LD. Hypersusceptibility and genetic problems in occupational medicine: a consensus report. *Journal of Occupational Medicine* 1973;15:564-573.

8. Mittman C, Lieberman J. Screening for alpha-1-antitrypsin deficiency. In: Ramot B, Adam A, Bonne B, Goodman R, Szeinberg A, eds. *Genetic polymorphisms and diseases in man.* New York: Academic Press, 1973:185-192.

9. Laurell CB, Eriksson S. Electrophoretic alpha-1-globulin pattern of serum in alpha-1-antitrypsin deficiency. *Scandinavian Journal of Clinical and Laboratory Investigation* 1963;15:132.

10. Tarkoff MP, Kueppers F, Miller WF. Pulmonary emphysema and alpha-1-antitrypsin deficiency. *American Journal of Medicine* 1968;45:220-228.

11. Talamo RC, Langley CE, Reed CE, Makino S. Alpha-1-antitrypsin: a variant with no detectable alpha-1-antitrypsin. *Science* 1973;181:70-71.

8. DETERMINING CONCLUSIONS AND RECOMMENDATIONS

The final task in preparing a health assessment is to determine conclusions about the health implications associated with the site and to prepare parallel recommendations. Completion of this task will fulfill the purposes of a health assessment stated in Chapter 2.

- Determine the health implications of the site.

- Address these implications by making recommendations for future environmental and health studies (if deemed necessary).

- Identify actions necessary to mitigate or prevent adverse health effects.

Recommendations should parallel conclusions drawn about the site, with every conclusion leading to one or more recommendations. It is imperative that this section be explicit and unambiguous and concisely state the findings of the health assessment. This section should logically follow from information presented in previous sections and should not introduce new data or information. Conclusions and resulting recommendations should address all community health concerns.

8.1. DETERMINING CONCLUSIONS

The health assessment conclusions should explicitly communicate the following:

- health implications of the site;

- community concerns; and

- instances of insufficient health and environmental information.

Additional conclusions for addressing specific health concerns or environmental pathways may also be needed.

8.1.1. Selection of the Public Health Hazard Category

The first conclusion of every health assessment identifies the level of public health hazard posed by the site. A health assessment should assign to the site one of the following five categories:

A. Urgent Public Health Hazard;
B. Public Health Hazard;
C. Indeterminate Public Health Hazard;
D. No Apparent Public Health Hazard; or
E. No Public Health Hazard.

These categories were selected to:

- characterize the degree of public health hazard at the site based on factors such as the existence of potential pathways of human exposure, the susceptibility of the exposed community, the comparison of expected human exposure levels to applicable health-based standards, and an evaluation of existing community-specific health outcome data.

- determine whether actions should be taken to reduce human exposure to hazardous substances from a site and whether additional information on human exposure and associated health risks is needed and should be acquired by conducting further environmental sampling or other health actions including epidemiologic studies, establishing a registry or a health surveillance program, or environmental health education.

- identify toxicologic data gaps for substance-specific and generic toxicologic issues. These data gaps will be considered by the Division of Toxicology in establishing research priorities and in developing Toxicological Profiles.

Table 8.1. contains criteria for selecting the appropriate site category. The site category is determined primarily by existing conditions at the site. Information on past exposures at the site will seldom be available. However, there may occasionally be information to indicate that past exposures to hazardous substances have affected human health. In some instances, these past exposures could have resulted in adverse health outcomes that have persisted to the present, even though the site has been remediated and exposures are no longer occurring. In order to acknowledge and be responsive to the persisting health impact of such past exposures, the site should be placed in Category A or B. In addition, if possible, ATSDR should make recommendations to mitigate the health impact of past exposures. They can include initiating medical monitoring or surveillance, establishing a registry, or implementing other appropriate actions.

Conditions at a site may be dramatically altered as the result of removal operations, remedial activities, or other intervention strategies. Conditions may also be altered as the result of unabated contaminant migration or changes in land use at the site or surrounding property. If such changes occur and if resources permit, an addendum to the health assessment can be prepared, and the site can be recategorized to reflect the altered circumstances.

In selecting the appropriate health hazard category, the assessor must consider the total body of information available for the site. Some of the important factors that must be weighed in the analysis include these:

- presence of completed or potential exposure pathways;

- on-site and off-site environmental contaminant concentrations;

- potential for multiple source exposures;

- contaminant interactions;

- presence of sensitive subpopulations;

- opportunity for acute or chronic exposures;

- nature of toxic effects associated with site contaminants;

- community-specific health outcome data;

- community health concerns; and

- presence of physical hazards.

Once particular populations are identified as being at risk for adverse health effects from site contamination, the health assessor should determine the course(s) of action needed to protect public health and to prevent human exposure. Those actions are reported in the Recommendations section of the health assessment report.

8.1.2. Conclusions About Insufficient Information

Gaps in environmental characterization data or the lack of sufficient human health information may have been identified during the health assessment process. These data gaps should be reported in the Conclusions section of the health assessment. Conclusions about insufficient information and the attendant potential health implications may be coupled in one statement.

8.1.3. Conclusions About Community Health Concerns

As discussed in Section 7.7., the health assessment should address all community health concerns using the criteria described in Chapter 7. Key findings from the evaluation of community health concerns should be presented in the Conclusions section of the health assessment. Such findings could include the plausibility of a specific concern and follow-up actions that are recommended to help address those concerns.

Table 8.1. Criteria and Actions for Levels of Public Health Hazard

CATEGORY A URGENT PUBLIC HEALTH HAZARD	CATEGORY B PUBLIC HEALTH HAZARD
This category is used for sites that pose an urgent public health hazard as the result of short-term exposures to hazardous substances.	*This category is used for sites that pose a public health hazard as the result of long-term exposures to hazardous substances.*

CATEGORY A — Criteria:

Evidence exists that exposures have occurred, are occurring, or are likely to occur in the future;

and

the estimated exposures are to a substance or substances at concentrations in the environment that, upon short-term exposures (less than 1 year), can cause adverse health effects to any segment of the receptor population. The adverse health effect can be the result of either carcinogenic or noncarcinogenic toxicity from a chemical exposure. For a noncarcinogenic toxic effect, the exposure exceeds an acute or intermediate minimal risk level (MRL) established in the ATSDR Toxicological Profiles or other comparable value;

and/or

community-specific health outcome data indicate that the site has had an adverse impact on human health that requires rapid intervention;

and/or

physical hazards at the site pose an imminent risk of physical injury.

ATSDR Actions:

ATSDR will expeditiously issue a health advisory that includes recommendations to mitigate the health risks posed by the site. The recommendations issued in the health advisory and/or health assessment should be consistent with the degree of hazard and temporal concerns posed by exposures to hazardous substances at the site.

Based on the degree of hazard posed by the site and the presence of sufficiently defined current, past, or future completed exposure pathways, the following public health actions can be recommended:

* biologic indicators of exposure study;
* biomedical testing;
* case study;
* disease and symptom prevalence study;

(Continued on next page)

CATEGORY B — Criteria:

Evidence exists that exposures have occurred, are occurring, or are likely to occur in the future;

and

the estimated exposures are to a substance or substances at concentrations in the environment that, upon long-term exposures (greater than 1 year), can cause adverse health effects to any segment of the receptor population. The adverse health effect can be the result of either carcinogenic or noncarcinogenic toxicity from a chemical exposure. For a noncarcinogenic toxic effect, the exposure exceeds a chronic MRL established in the ATSDR Toxicological Profiles or other comparable value;

and/or

community-specific health outcome data indicate that the site has had an adverse impact on human health that requires intervention.

ATSDR Actions:

ATSDR will make recommendations in the health assessment to mitigate the health risks posed by the site. The recommendations issued in the health assessment should be consistent with the degree of hazard and temporal concerns posed by exposures to hazardous substances at the site.

Based on the degree of hazard posed by the site and the presence of sufficiently defined current, past, or future completed exposure pathways, the following public health actions can be recommended:

* biologic indicators of exposure study;
* biomedical testing;
* case study;
* disease and symptom prevalence study;
* community health investigation;

(Continued on next page)

Table 8.1. Continued

CATEGORY A URGENT PUBLIC HEALTH HAZARD (continued)	CATEGORY B PUBLIC HEALTH HAZARD (continued)
• community health investigation; • registries; • site-specific surveillance; • voluntary residents tracking system; • cluster investigation; • health statistics review; • health professional education; • community health education; and/or • substance-specific applied research.	• registries; • site-specific surveillance; • voluntary residents tracking system; • cluster investigation; • health statistics review; • health professional education; • community health education; and/or • substance-specific applied research.

Table 8.1. Continued

CATEGORY C INDETERMINATE PUBLIC HEALTH HAZARD	CATEGORY D NO APPARENT PUBLIC HEALTH HAZARD
This category is used for sites with incomplete information.	*This category is used for sites where human exposure to contaminated media is occurring or has occurred in the past, but the exposure is below a level of health hazard.*
Criteria:	**Criteria:**
The limited available data do not indicate that humans are being or have been exposed to levels of contamination that would be expected to cause adverse health effects. However, data or information are not available for all environmental media to which humans may be exposed; **and** there are insufficient or no community-specific health outcome data to indicate that the site has had an adverse impact on human health.	Exposures do not exceed an ATSDR chronic MRL or other comparable value; **and** data are available for all environmental media to which humans are being exposed; **and** there are no community-specific health outcome data to indicate that the site has had an adverse impact on human health.
ATSDR Actions:	**ATSDR Actions:**
ATSDR will make recommendations in the health assessment to identify the data or information needed to adequately assess the public health risks posed by the site. Public health actions recommended in this category will depend on the hazard potential of the site, specifically as it relates to the potential for human exposure of public health concern. If the potential for exposure is high, initial health actions aimed at determining the population with the greatest risk of exposure can be recommended. Such health actions include: • community health investigation; • health statistics review; • cluster investigation; and • symptom and disease prevalence study. If the population of concern can be determined through these or other actions, any of the remaining follow-up health activities listed under categories A and B may be recommended. In addition, if data become available suggesting that human exposure to hazardous substances at levels of public health concern is occurring or has occurred in the past, ATSDR will reevaluate the need for any followup.	If appropriate, ATSDR will make recommendations for monitoring or other removal and/or remedial actions needed to ensure that humans are not exposed to significant concentrations of hazardous substances in the future. The following health actions, which may be recommended in this category, are based on information indicating that no human exposure is occurring or has occurred in the past to hazardous substances at levels of public health concern. The following health actions are recommended for sites in this category: • community health education; • health professional education; • community health investigation; and • voluntary residents tracking system. However, if data become available suggesting that human exposure to hazardous substances at levels of public health concern is occurring, or has occurred in the past, ATSDR will reevaluate the need for any followup.

Table 8.1. Continued

CATEGORY E
NO PUBLIC HEALTH HAZARD

This category is used for sites that do not pose a public health hazard.

Criteria:

There is no evidence of current or past human exposure to contaminated media;

and

future exposures to contaminated media are not likely to occur;

and

there are no community-specific health outcome data to indicate that the site has had an adverse impact on human health.

ATSDR Actions:

No public health actions are recommended at this time because no human exposure is occurring, has occurred in the past, or is likely to occur in the future that may be of public health concern.

8.1.4. Conclusions About Exposure Pathways

Conclusions about exposure pathways may be appropriate when the environmental pathways may be affected by remediation or when the pathways may contribute to potential future exposures. Conclusions about "hot spots" should also be included when these areas have significant health implications different from those of the rest of the site.

8.2. DETERMINING RECOMMENDATIONS

After reaching conclusions about the site, the assessor should develop recommendations for:

1) implementing actions to protect public health;

2) obtaining additional health information;

3) conducting public health actions (see public health action plan); and

4) obtaining additional site-characterization information.

Recommendations are based on the sites's hazard category (Table 8.2.) and correspond directly to the conclusions identified previously. Inherent in every recommendation should be the urgency with which the recommendation needs to be addressed. This measure of urgency will indicate the gravity of the attendant conclusion and establish priorities for responding to the recommendation.

Table 8.2. Conclusion Categories and Corresponding Recommendations and Public Health Actions

CONCLUSION CATEGORIES	EXAMPLE RECOMMENDATION	EXAMPLE PUBLIC HEALTH ACTION
A: Urgent Public Health Hazard	Dissociate persons from exposure to hazardous substances at the site	• Public health advisory • Conduct blood lead testing to determine if an immediate public health threat exists for the community living near the site
B: Public Health Hazard	Restrict access to the contaminated areas	• Conduct biomedical indicators of exposure study or biological testing to validate exposures
C: Indeterminate Public Health Hazard	Conduct private well survey to better identify exposed population	• Conduct community health investigation to characterize population at risk
D: No Apparent Public Health Hazard	Conduct periodic monitoring of all area private wells to ensure that no exposure is occurring to hazardous substances at levels of public health concern	• Conduct community health education to assist residents in understanding their potential for exposure
E: No Public Health Hazard	None	• None

8.2.1. Recommendations for Protecting Public Health

For a site that poses an urgent public health hazard, ATSDR may issue a health advisory, which is sent to the EPA and to the appropriate state public health agency. A health advisory should be considered whenever chemical contamination or physical hazards associated with a site necessitate an expeditious response to protect public health. The health advisory recommends measures to be taken to reduce exposures and to eliminate or substantially mitigate the significant risk to human health.

Recommendations for protecting public health are also included in the health assessment. The assessors may make recommendations to EPA for removal or remedial measures at the site to prevent further exposure, or they may recommend additional studies of the populations to better define the magnitude of exposure and the resulting health effects. Recommendations should address the following routes of exposure:

Human exposure from contaminants in water. When the assessor determines that humans are receiving unacceptable contaminant exposure through water sources, several types of action may be recommended: 1) provide an alternate water source; 2) close the public water supply or initiate treatment to remove contaminants; 3) evacuate properties that will be or are being adversely affected by surface-water releases until such releases are properly controlled; 4) restrict all public access to a site or contamination zone; 5) continue monitoring contaminant migration in groundwater; and 6) recommend that EPA give the site a higher priority.

Human exposure through soil. When contaminated soil poses a health threat to on-site or near-site populations, the assessor may recommend to EPA, state environmental or health agencies, or other appropriate authority(ies) that site access be restricted. The assessor might also recommend that appropriate removal or remedial actions be implemented to prevent contact with on-site contaminated soil or off-site transport of soil by water or wind erosion.

Human exposure through the food chain. If the assessment shows human exposure to food contaminants at levels that pose a health threat, EPA, the Food and Drug Administration (FDA), or other appropriate authority(ies) should be advised to prohibit local consumption or transportation of contaminated foodstuffs to other locations for consumption. If fish, game animals, or consumable wild plants are contaminated, the proper authorities should issue appropriate advisories.

If measurements show contaminant levels that are elevated, but below levels of human health concern, the appropriate authorities should issue advisories that consumption of these contaminated foodstuffs, while not in and of itself a definitive or extraordinary human health threat, may pose a human health threat in conjunction with exposure to similar contaminants in other foodstuffs or by other routes (such as occupational or household exposure). That type of advisory should be distributed near the site and wherever transported contaminated plants and animals are being consumed.

Safety actions. Recommendations on actions that should be taken to protect public health, including remedial worker safety, should be included in the Recommendations section of the health assessment.

8.2.2. Recommendations for Public Health Actions

The final recommendation in every health assessment should pertain to follow-up health actions in populations living near hazardous waste sites. A variety of health actions can be undertaken as followup to a health assessment. There are three major types of follow-up actions: environmental health education, health studies, and substance-specific applied research.

To coordinate this process and to guide the health assessor, an agency-wide panel was established within ATSDR. The Health Activities Recommendation Panel (HARP) was established to:

- recommend follow-up health activities for all draft health assessments, draft public health advisories, and draft health consultations addressing health-related actions;

- track these recommendations; and

- ensure implementation of recommended activities.

Table 8.2.2. presents three questions that are used to guide the recommendation process. Each question is accompanied by a list of follow-up actions that may be considered if there is an affirmative answer to the corresponding question.

If the available information is insufficient to answer the questions in Table 8.2.2., the health assessor should contact appropriate federal, state, and local agencies to acquire the needed information.

The possible follow-up health actions listed in Table 8.2.2. are linked to the degree of hazard a site poses and are listed as options for ATSDR actions under each conclusion category (Table 8.1.).

Table 8.3. provides a list of general criteria for conducting each of the 13 follow-up health actions.

8.2.3. Recommendations for Additional Environmental Information

At times, site information is not available or is insufficient and cannot be used to adequately characterize site environmental characteristics, types and extent of contamination, and locations of potential receptor populations. In those cases, brief explicit recommendations should be made in the Recommendations section of the health assessment to specifically indicate the information required and the means by which that information should be obtained.

8.2.4. Outlining Public Health Actions

Based on the recommendations presented in the health assessment, the health assessor is now able to identify health actions that have been undertaken and those that are planned to address the public health impact of a specific site. These actions can vary from health investigations in the community living near the site to environmental characterization activities to better identify populations at risk of exposure. This information is outlined in the subsection "Public Health Actions" in the Recommendations section. The purpose of this subsection is to clearly delineate in the health assessment the site-specific health agenda designed to mitigate or prevent adverse human health effects resulting from exposure to hazardous substances that may be associated with a particular site.

Table 8.2.2. Questions used to guide the recommendation process

Question 1: **Does the potentially exposed population need assistance in understanding its potential for exposure or in assessing adverse health occurrences in its community?**

If **YES**, see specific conditions and criteria for:

Community Health Education
Community Health Investigation
Health Professional Education
Voluntary Residents Tracking System

Question 2: **Is human exposure believed to be occurring or to have occurred in the past because of human interaction with a pathway of exposure known to be contaminated by a hazardous substance?**

If **YES**, see specific conditions and criteria for:

Biological Indicators of Exposure Study
Biomedical Testing
Case Study
Disease- and Symptom-Prevalence Study
Community Health Investigation
Registries
Site-Specific Surveillance
Substance-Specific Applied Research
Voluntary Residents Tracking System

Question 3: **Is there an indication or allegation that adverse health conditions are occurring in the population that may be related to exposure to hazardous substances in the environment?**

If **YES**, see specific conditions and criteria for:

Biomedical Testing
Case Study
Cluster Investigation
Community Health Education
Disease- and Symptom-Prevalence Study
Health Professional Education
Health Statistics Review
Registries
Site-Specific Applied Research
Substance-Specific Applied Research

Developing the Public Health Action Subsection

The Public Health Action subsection consists of two parts: "actions undertaken" and "actions planned."

Actions undertaken

First, the health assessor should indicate health actions undertaken to respond to recommendations outlined in the health assessment. For example, if a previously recommended private well survey was conducted, that information should be provided in this subsection. The actions may have been carried out by one of the various agencies involved, including ATSDR, EPA, and state and local health and environmental departments.

Actions planned

Next, the health assessor should delineate public health actions that will be carried out based on the recommendations presented in the health assessment. The health actions planned will generally consist of two types: those recommended by the ATSDR Health Activities Recommendation Panel (HARP) and those committed to by agencies other than

ATSDR. For example, the health assessor could state that ATSDR, in cooperation with a state health department, will conduct blood testing in children living near the site, or that EPA will restrict site access. In addition to identifying the agency that will undertake the activities outlined in a specific recommendation, it is helpful to indicate when, if possible, the activities will take place.

To develop and implement an effective site-specific health agenda, it is important to identify the organizations and agencies most capable of carrying out the activities outlined in each recommendation. The Public Health Actions subsection is developed after all recommendations are outlined in the health assessment, including those made by HARP. During the initial release of the health assessment, agencies involved with the site other than ATSDR will be asked to identify actions they have undertaken or are able to commit to based on the recommendations outlined in the health assessment. In this phase of the health assessment, health assessors are encouraged to contact those organizations to discuss pertinent recommendations. Once developed, the site-specific health agenda will be made available to the public for comment.

Table 8.3. Criteria For Determining Follow-up Health Activities
Note: For each activity, **all** criteria must be met.

Biological Indicators of Exposure Study

1. Human exposure is believed to be occurring or may have occurred because of human interaction (such as direct contact, inhalation, or ingestion) with a pathway of exposure known to be contaminated by a hazardous substance(s).

2. Persons potentially exposed along the pathway can be identified and located for testing.

3. Adequate quality-controlled and sensitive laboratory test(s) is (are) available to detect the presence of the hazardous substance(s), its (their) metabolite(s), or other biologic marker(s) known to be closely associated with exposure and measurable in some biologic tissue or fluid.

4. Previous experience and scientific knowledge are inadequate or insufficient to predict whether individual biologic uptake of hazardous substances or illness will occur as a result of the environmental conditions present at the site.

Biomedical Testing

1. Human exposure is believed to be occurring or may have occurred in the past because of human interaction (such as direct contact, inhalation, or ingestion) with a pathway of exposure known to be contaminated by a hazardous substance(s).

2. Persons potentially exposed along the pathway can be identified and located for testing.

3. Health effects under study are biologically plausible and may be postulated to be caused by the exposure at the concentrations observed.

4. Standard medical test(s) or adequate quality- controlled and sensitive laboratory test(s) is (are) available to detect the biological effects being evaluated.

Case Study

1. Individuals are located along potential pathway(s) of exposure associated with a hazardous waste site.

2. Exposure of those persons to hazardous waste has been documented, or a reasonable concern exists for the potential of an as-yet-unidentified route of exposure.

3. A reasonable concern for adverse health effects has been hypothesized for individuals at potential risk.

4. Case information about adverse health effects or exposure to hazardous substances can be obtained for comparison to develop a hypothesis about the relationship between the exposure to hazardous substances and adverse health effects.

Cluster Investigation

1. A human population is in the vicinity of a hazardous waste site.

2. Human exposure to hazardous substances(s) has been documented, or a reasonable concern exists for the potential of an as-yet-undefined route of exposure.

Table 8.3. Continued

3. A reasonable concern for public health exists as a result of reports of disease in the community.

4. Information can be located or collected to verify the disease and document the geographic and temporal occurrence of the cases.

5. Biologic plausibility supports a relationship between hazardous substance(s) at the site and disease being reported.

Community Health Education

1. A human population lives/works along potential pathway(s) of exposure associated with a hazardous waste site.

2. Concern for public health exists as a result of reports about exposures and/or reports of disease in the community.

3. Human exposure to hazardous substance(s) may have occurred or may be occurring.

Community Health Investigation

1. A human population lives/works along potential pathway(s) of exposure associated with a hazardous waste site.

2. Human exposure is believed to be occurring or may have occurred because of human interaction (such as direct contact, inhalation, or ingestion) with an exposure pathway known to be contaminated by hazardous substance(s).

3. Concern for public health exists as a result of reports of disease in the involved population.

4. Information is available on relevant health outcome data for the involved population.

5. Community health concerns exist related to the site.

Disease- and Symptom-Prevalence Study

1. A human population lives/works in the vicinity of a hazardous waste site.

2. Exposure of persons to hazardous substance(s) has been documented, or a reasonable concern exists for the potential of an as-yet-unidentified route of exposure.

3. A reasonable concern for adverse health effects has been hypothesized for individuals at potential risk.

4. Information about adverse health effects or exposure to hazardous substances can be obtained for comparison and to develop a hypothesis about the relationship between the exposure to hazardous substances and adverse health effects.

Health Professional Education

1. A human population lives/works along potential pathway(s) of exposure associated with a hazardous waste site.

2. Human exposure to hazardous substance(s) has been documented, or a reasonable concern exists for the potential of an as-yet-unidentified route of exposure.

3. A reasonable concern for public health exists as a result of reports of disease in the community. One or both of the following conditions should be present:

Table 8.3. Continued

a. There are community concerns about the site, and there is a lack of information available from local public health professionals.

b. Citizens have expressed concern that local, private medical practitioners lack information on the potential health effects of site hazards.

4. A specific request has been received from individuals, health care providers, special interest groups, industry, academia, or government agencies for health professional education related to an NPL site, a non-NPL site or facility, an emergency response site, or other site or facility.

Health Statistics Review

1. A human population is located in the vicinity of a hazardous waste site.

2. Exposure of persons to hazardous substance(s) has been documented, or concern exists for the potential of an as-yet-unidentified route of exposure.

3. An indication or allegation exists that adverse health conditions that may be related to exposure to hazardous substances are occurring in the population .

4. Data on relevant health outcomes for the involved population are available. Data are accessible and can provide health outcome information applicable to the population, or data manipulation can yield relevant health outcome information about the population (if data were not collected in a fashion that is readily applicable to the population).

Registries

1. The contaminant is one for which a subregistry has been established.

2. The site is an NPL site, a non-NPL site, an emergency response site, or other site identified by individuals, special interest groups, industry, academia, or other government agencies as involving the release of hazardous substance(s) into the environment.

3. The site fits within the general guidelines considered in site selection for a registry as established in *Policies and Procedures for Establishing a National Registry of Persons Exposed to Hazardous Substances* (ATSDR,1988)—that is, human exposure has been documented; the size of the potentially exposed population is acceptable; the presence or absence of reported health problems has been verified; and the community is interested in participating.

Site-Specific Surveillance

1. Human exposure is believed to be occurring or may have occurred in the past at a level and in such a manner that is biologically relevant because of human interaction (such as direct contact, inhalation, or ingestion) with a pathway of exposure known to be contaminated by hazardous substance(s).

2. Persons potentially exposed along this pathway can be identified and located for testing and/or followup.

3. The outcome to be measured is biologically plausible and relevant.

4. Adequate quality-controlled and sensitive laboratory test(s) is (are) available to detect the presence of hazardous substance(s), its (their) metabolite(s), or other biologic marker(s) known to be closely associated with exposure and measurable in some biologic tissue or fluid, or there is a measurable and sensitive health outcome that can be identified through existing data sources, such as medical records.

Table 8.3. Continued

5. Previous experience and scientific knowledge are inadequate or insufficient to predict whether biologic uptake of hazardous substances or illness will occur under the environmental conditions present at the site.

6. The identified cohort of potentially exposed persons is willing to participate in a longitudinal survey.

Substance-Specific Applied Research

1. A contaminant of concern at the site is not the subject of an ATSDR Toxicological Profile.

2. A Toxicological Profile exists for a contaminant of concern at the site; however, the Profile needs updating.

3. A current Toxicological Profile exists for a contaminant of concern at the site; however, information required for this site is listed as a data need.

4. A current Toxicological Profile exists for a contaminant of concern at the site; however, information required for this site is not addressed in the Profile.

Voluntary Residents Tracking System

1. A human population lives/works along potential pathway(s) of exposure associated with a hazardous waste site.

2. Human exposure to hazardous substance(s) has been documented, or a reasonable concern exists for the potential of an as-yet-unidentified route of exposure.

3. A reasonable concern about public health exists as a result of reports of disease in the community.

4. Health communication or future health studies can be developed that require creation of a list of persons potentially exposed to hazardous substances from the site.

9. HEALTH ASSESSMENT FORMAT

Once the site visit has been made, the available documents have been reviewed, and the evaluation has been completed, the health assessor can proceed with writing the health assessment. This section describes the format the health assessment should follow and provides guidance on what should be included in each section of the health assessment report. All health assessments (including preliminary health assessments and petitioned health assessments) should follow the same format, even though the focus and amount of detail may vary.

Good communication skills are vital when conveying public health concerns in the health assessment. Public health practice involves not only using the technical skills of specific health disciplines but also communicating information the public can understand and use to make responsible health decisions. The assessor must, therefore, be capable of explaining difficult scientific concepts, using understandable environmental and medical terminology and a minimum of jargon.

The assessor must relate to the public. The health assessment should be written using simple, direct statements that explain public health issues. The health assessor should find appropriate ways to personalize site information to provide relevant answers to community and personal health issues.

The health assessment focuses on seven primary kinds of information: history of regulatory involvement, community health concerns, site visit observations and interviews, contaminant assessment, pathways analyses, toxicologic information, and relevant health outcome databases. The health assessment ends with conclusions about health hazards posed by the site and recommendations to end or reduce exposure; to characterize the site further; and to suggest follow-up health

Table 9.1. Health Assessment Format

Summary

Background
A. Site Description and History
B. Site Visit
C. Demographics, Land Use, and Natural Resources Use
D. Health Outcome Data

Community Health Concerns

Environmental Contamination and Other Hazards
A. On-Site Contamination
B. Off-Site Contamination
C. Quality Assurance and Quality Control
D. Physical and Other Hazards

Pathways Analyses
A. Completed Exposure Pathways
B. Potential Exposure Pathways

Public Health Implications
A. Toxicologic Evaluation
B. Health Outcome Data Evaluation
C. Community Health Concerns Evaluation

Conclusions

Recommendations (Public Health Actions)

Preparers of Report

References

Appendices

activities. Table 9.1. outlines the health assessment format. The content of each section in the health assessment is described in the following pages.

Example 1.

9.1. SUMMARY

The Summary should be a concise synopsis of the health assessment and should state the following:

- site location, including nearest city or town, county, and state;

- contaminants that drive the selection of the conclusion categories "urgent public health hazard" and "public health hazard;"

- exposed populations;

- community health concerns;

- populations at special risk;

- completed exposure pathways;

- potential exposure pathways;

- health effects in exposed populations;

- conclusions about the public health category assigned to the site;

- urgency of the site's public health implications; and

- recommendations.

Example 1. provides a sample Summary.

SUMMARY

The Canyon Waste Dump site is a former solvents recovery operation near Canyonville, Calizona. Site-related contamination poses an urgent public health hazard because acute, intermediate, and chronic exposure from eating mercury-contaminated fish from the Santa River could cause adverse health effects including kidney damage, damage to fetuses, and damage to the brain. In addition, long-term exposure to trichloroethene in private wells along Canyon Road could slightly increase the risk of cancer.

A public health hazard exists for young children who lived or are living along Canyon Wash. Because of exposure to mercury from ingesting soil and sediment contaminated by the site, damage could occur to the kidneys and the brain. While lead in the Canyonville Public Water supply is unrelated to the site, it still remains a public health hazard. Lead-contaminated water can cause adverse health effects in infants and young children, resulting in decreased IQ scores, growth retardation, and hearing problems.

Citizens raised several questions about learning disabilities, birth defects, and other health effects believed to have resulted from exposure to site-related contaminants. Detailed answers to those questions appear in the Public Health Implications section of the health assessment.

The Agency for Toxic Substances and Disease Registry (ATSDR) will be issuing a health advisory for the fish exposure pathway. ATSDR has also made recommendations to 1) reduce and prevent exposure to contaminants, 2) better characterize the site, 3) implement health followup and other activities, 4) address community health concerns, and 5) identify toxicologic data gaps.

9.2. BACKGROUND

The Background section provides a historical perspective on the site, and describes its operations, current status, and the surrounding community.

9.2.1. Site Description and History

This subsection should begin with a brief statement of the problems known or suspected at the site (e.g., contaminated groundwater resulting from mismanaged industrial wastes, or contamination resulting from an abandoned landfill). A brief description of local, state, and federal government involvement at the site, particularly ATSDR's past involvement, should follow. How and why the health assessment was initiated should also be explained (e.g., the site is an NPL site, a RCRA site, or a petitioned site). The health assessor should note pertinent regulatory activities, such as emergency removal or response actions taken at the site, and the site's current regulatory status (e.g., a site investigation report was recently completed; a final Feasibility Study was issued in May 1989; or this health assessment is being performed in connection with a forthcoming Record of Decision). Example 2. illustrates how this information may be reported in a health assessment.

This subsection next should focus on physical, historical, and operational details of the site. The discussion should be concise and should present only facts necessary to evaluate exposure pathways and health implications associated with the site. Points to be covered include:

• site location (size, area, and physical makeup);

• site features (surface features, e.g., ravines, ditches, and surface cover);

Example 2.

Because of an August 1985 permit application for replacement of two Canyon Road bridges, one across Santa River and the other across Canyon Creek, the Calizona Department of Natural Resources (CDNR) and the U.S. Environmental Protection Agency (EPA) discovered that sediments in the Santa River contained high mercury levels. The Calizona Department of Transportation, with the assistance of CDNR, collected sediment samples from Santa River for the permit. High mercury levels in the sediment and possible uptake in fish led EPA and the CDNR to contact the U.S. Fish and Wildlife Service for technical assistance in collecting fish samples. Because the Santa River fish samples contained high mercury concentrations, the CDHS issued a health advisory banning consumption of fish from the Santa River five miles up- and downstream of Canyon Road Bridge.

In June 1987, EPA conducted a site assessment to determine the extent of contamination on and off the site. EPA collected environmental samples of waste materials, soil, sediments, surface water, groundwater, soil gas, and air. EPA evaluated those data and conducted a clean-up action in June 1989. They removed 5,000 drums, 5 tanks, and 5 million gallons of liquid wastes and sludge from the waste pit to a state-approved hazardous waste site. EPA also sampled three private wells near the site and provided bottled water to families drinking from the two wells found to be contaminated.

EPA is developing a work plan for a Remedial Investigation/Feasibility Study (RI/FS) which will further characterize the extent of contamination and evaluate alternatives for cleaning up the site. EPA expects to complete the cleanup by June 1993.

- site history in chronological order (origins of the site, changes in size, and industrial and commercial activities);

- operational activities carried out at the site when specific processes or operations characterize the waste present;

- current site owner;

- current status; and

- access to site (e.g., fences, gates, security).

The detail and length of this section of the report will vary according to the site. This subsection, however, should provide only pertinent information about the site that helps the reader to understand the Pathways Analyses and the Public Health Implications sections of the health assessment. Example 3. provides a sample of a site description and history.

9.2.2. Site Visit

Observations from the site visit are an important part of the health assessment. When discussing the site visit, the health assessor should include:

- the names of each ATSDR staffer, ATSDR subcontractor, or state cooperative agreement staffer who conducted the site visit;

- the names of other agencies and organizations participating in the site visit, but not the names of individuals and private citizens;

- the date and time spent visiting the site and surrounding areas;

- observations made during the site visit (to differentiate between observations summarized in other documents); and

Example 3.

The 69-acre Canyon Waste Dump site, a former solvents recovery operation, lies at the head of Horseshoe Canyon, about 1.5 miles south of Canyonville, Calizona, in Canyon County. Figure 1 shows the important features on and near the site. Surrounded by a fence with a locked gate, the site contains in its center a 23-acre waste pit with standing water. No buildings or other man-made features exist on the remaining 46 acres of bare ground. A wash, which drains the head wall of the canyon and the subdivisions above the canyon, has eroded portions of the site inside the east perimeter fence. No part of the site lies within a corporate boundary; the nearest corporate boundary belongs to the city of Canyonville.

Following the purchase of idle land in the canyon, Waste Systems Ltd. began activity at the site in 1965. The state of Calizona and the County issued permits for a storage pit to be used for solvents recovery operations. The pit's design incorporated a 10-foot clay dike and a pit liner. For the pit liner, the facility used an existing shaley-clay layer beneath the pit. Although the site owner constructed some buildings, no recovery operations occurred on the site. No records exist about wastes stored on the site.

Storage operations continued until September 1975, when the State canceled the facility's permit because of improper dike maintenance. Dike failure in April 1974 and March 1975 resulted in two known spills from the waste pit into Canyon Wash. Despite claims by Waste Systems Ltd. that the spills had been cleaned up, CDHS reported that it had documented evidence that no cleanup occurred (personal communication, CDHS). In September 1975, Waste Systems Ltd. stopped operations, dismantled and hauled off on-site buildings and equipment, and sold the site to the County for development of a landfill. Although the County has maintained the dike around the pit since 1976, the site has not been used for any purpose and is considered inactive.

- other site-specific information, such as:

  - visual evidence of contamination,

  - evidence of authorized or unauthorized site activities,

  - observations of populations or subpopulations at risk,

- remediation or mitigation activities not discussed in available documents,

- security and site accessibility by the public,

- environmental monitoring requested or performed by ATSDR,

Example 4.

Dr. Robert Jones and Ms. Elizabeth Appleton with CDHS and Mrs. Louise Hall, the ATSDR regional representative, visited the site area September 4-6, 1990. On the first day, we inspected the site and made the following observations:

- The gate was unlocked, and the "No Trespassing" signs were lying on the ground.
- Portions of the fence in the northwestern corner were missing .
- The ground was bare of vegetation.
- The pit contained about 2 feet of standing water and three visible drums.
- Eroded gullies up to 10 feet deep lined the entire eastern side of the site near the perimeter fence.
- No persons or buildings were seen on or near the site.
- Off-road bike tracks were visible throughout the site.
- Four piles of household garbage were found just inside the gate.

Off site, we saw a fenced power substation 1,600 feet from the northeastern corner of the site and many single-family residences with visible wells along Canyon Road. The homes were built in the 1950s (personal communication, City Planner of Canyonville, 1990). The closest residence, a single-family home, lies 1,500 feet north on Canyon Road. In addition, we noticed two subdivisions near the site: Canyon Spring Subdivision, 4,000 feet north of the site; and Canyon Top Heights, 2,000 feet south. Canyon Pointe, an undeveloped subdivision with a paved road, lies just northwest of Canyon Top Heights.

Except for the Canyon Spring Subdivison, land surrounding the site has little or no vegetation. Residential yards along Canyon Road also lack vegetation. The City Planner told us that the county paved Canyon Road from Canyonville to the Canyon Spring Subdivision in 1983, the year the subdivision was built. The rest of Canyon Road between the subdivision and the site remains unpaved. The City Planner also told us that the Canyon Top Heights subdivision was developed between 1970 and 1975. Further down the canyon, near the confluence of Canyon Wash and Canyon Creek and about 1.5 miles from the site, we observed Tots Nursery and Canyonville Elementary School. Those facilities have existed since the 1960s (personal communication, City Planner of Canyonville, 1990).

We obtained information about the area from several local officials: the Canyonville city planner, the Tots Nursery manager, the Canyonville Elementary School principal, the Canyon County Public Works and Roads director, the Canyonville Water Supply superintendent, the Canyon Spring Public Water Supply manager, the Coalition to Save Canyonville chairperson, the Canyonville Hospital director, and the Canyon County Health Department director. We have incorporated information obtained from those persons into appropriate sections of this health assessment.

- land uses,

- local human activities,

- nature of the terrain and visible environmental transport routes, and

- physical hazards at the site.

When warranted, a copy of a site summary form (Appendix L) may be included as an appendix to the health assessment. Example 4. provides a description of some, but not all, of the components of a site visit.

9.2.3. Demographics, Land Use, and Natural Resource Use

Once the important background details have been presented, the assessor should describe characteristics of populations on or near the site and the use of local land and resources. The text should state the total number of exposed and potentially exposed persons and indicate the characteristics of this population. Population estimates should include persons exposed in the past and currently, as well as those at risk for future exposures. Demographic information may include discussions of specific population groups surrounding the site (e.g., residential, commercial, and occupational populations). If warranted, details about the size; exact location; age distribution; and socioeconomic, genetic, and ethnic makeup of populations on and near the site should be discussed based on available information. Ethnic and socioeconomic background information is essential for a full understanding of the health threat a site poses to specific subpopulations. Populations that may be at special risk from exposure to the site, such as children, pregnant women, and elderly and infirm persons, should receive special note. Example 5. is a sample of information on demographics.

The land- and resource-use information should describe only relevant activities on or around the site, particularly those that are related to points of exposure. Information should also be provided about distances to

Example 5.

Approximately 100,000 persons live in Canyon County, Calizona. The population of Greater Canyonville lies within Horseshoe Canyon. According to the city planner and the CCHD Well Survey, 36,879 persons live in Greater Canyonville: 36,029 persons and 10,000 homes in the City of Canyonville; 200 persons and 56 homes along Canyon Road south of Canyonville; 400 persons and 111 homes in Canyon Spring Subdivision; and 250 persons and 70 homes in Canyon Top Heights.

The racial makeup of Greater Canyonville is 80 percent white and 20 percent black.* Of the two groups, 14 percent are of Spanish origin (2 percent white Hispanic and 12 percent black Hispanic). The percentage of black and Hispanic persons along Canyon Road is higher. Those populations, except for the newly developed Canyon Spring Subdivision, have remained constant since the 1950s. The Canyonville homes represent moderate-income housing, and the Canyon Road homes south of Canyonville and the Canyonville Spring Subdivision homes represent low-income housing. The Canyon Top Heights homes represent high-income housing (personal communication, City Planner for Canyonville).

*as reported in 1990 data from the U.S. Census Bureau

the nearest residences and other land use areas where receptor populations exist or may exist in the future:

- groundwater use (e.g., residential wells, public water supply wells, industrial or production wells),

- surface water use (e.g., water supply intakes, commercial and sport fishing areas, boating areas, swimming areas, and aquaculture),

- terrain use (e.g., recreational, park, and refuge areas used for camping, hunting, hiking, biking, birding, and sporting, etc.),

- agricultural use (e.g., crops, orchards, gardens, feedlots, pasture, dairy, beehives),

- health care and educational use (e.g., hospitals, nursing homes, day care, schools), and

- residential, commercial, and industrial land use.

In addition, the proposed future land use should be described and the text should note the distance between the site and the land proposed for development. Example 6 presents some of these concepts.

In many cases, adequate information on demographics and land and resource use is missing from materials provided to ATSDR. However, such information can usually be obtained from state and local authorities, and every effort should be made to include it. When demographic or land-use information is lacking or deficient, the assessor should evaluate whether such information is needed to accurately assess the site's public health implications. If necessary, a statement may be included explaining the limitations posed by the lack of data. Additional demographic or land and resource use information needed to complete the health assessment should be discussed in the Conclusions and Recommendations sections.

Example 6.

Canyon County has zoned the site for industrial use only. This zoning extends 1,000 feet north of the site. Other land near the site has not been zoned because it is too steep for development. The county has zoned "residential" the remaining areas in Horseshoe Canyon, except for 1) the power substation north of the site, 2) the commercial areas within Canyonville, and 3) an agricultural area east of Canyonville on both sides of the Santa River. A 50-acre vineyard lies about 1,000 feet east of Canyonville Elementary; other vineyards exist on both sides of the Santa River. No other agricultural activities occur in the canyon. The only planned change in zoning restrictions is development of a commercial area west of and adjacent to Canyonville.

Horseshoe Canyon residents use groundwater from either private wells or from public water supplies. For drinking and other household uses, residents along Canyon Road and Canyon Top Heights have private wells that draw water from a deep confined aquifer called the Canyon Sandstone. An aquifer consists of rock or rock materials that are sufficiently permeable to conduct groundwater and to yield sufficient quantities of water to wells or springs. A confined aquifer is an aquifer that is bounded above and below by impermeable materials. The Canyon Spring Public Water Supply, which serves about 36,029 persons, uses a shallow water table aquifer or unconfined aquifer. The manager of the Canyonville Water Supply informed us that it serves Tots Nursery and Canyonville Elementary and that the Santa River recharges its well.

9.2.4. Health Outcome Data

This subsection should document the health outcome databases selected for evaluation in the health assessment. Health outcome databases should be selected using three criteria: 1) the health outcome is plausibly related to a contaminant in a completed exposure pathway; 2) the health outcome is relevant to community health concerns; and 3) the database contains information on a health outcome identified in 1 or 2. The following information should be included:

- the complete name of the health outcome database,

- its source (state or local health department), and

- its relationship to exposed populations.

Example 7.

Using state health databases, it may be possible to determine if there is a higher than expected number of certain health effects in Greater Canyonville. This section identifies the relevant available databases; they are evaluated in the Public Health Implications section. While cancer may be a plausible health outcome from long-term exposure to at least one site contaminant, no cancer registries exist. Please refer to the Toxicologic Evaluation subsection of the Public Health Implications section for more information on cancer.

CDHS maintains two health databases relevant to this assessment: an infant mortality database and a birth defects registry. The infant mortality database gathers information such as geographic location, race, age, cause of death, etc., on infants ages 1 to 11 months. That information is available, for the years 1966 to 1988, for Greater Canyonville, Canyon County, and the state of Calizona.

Results of the evaluation of these databases should not be included in this subsection. Rather, the results should be included in the Health Outcome Data Evaluation subsection of the Public Health Implications. Example 7. provides a sample description of health outcome data.

9.3. COMMUNITY HEALTH CONCERNS

Addressing the health questions of the residents associated with a site is central to ATSDR's overall mission and to the purposes of the health assessment. The nature and degree of residents' health concerns will vary from site to site. However, addressing the health concerns of the community is crucial if the health assessment is to satisfy its purpose of helping the public and health professionals understand the risks posed by a site.

This section should objectively report questions raised either in written documents, or during site visits, or in meetings with residents, community groups, or elected officials. When such meetings occur, a brief discussion of the meeting and the concerns raised should be included. When possible, the number of persons who have concerns should be put into perspective with the number of persons who attend a public meeting or who have been informed about the opportunity to voice concerns. When several community groups are involved, each group and its respective concerns should be identified. The assessor should describe the community health concerns and, if raised, suspected exposures and health effects.

The assessor should remember that perspectives on risk and judgments about the site's health implications may be personal as well as professional, both for the

Example 8.

CCHD determined community health concerns by asking questions during its well survey. After our site visit, CDHS and CCHD convened a public availability session to hear the community's health concerns about the site. Of the 100 persons who attended, 20 voiced concern. During the well survey, the public meeting, and conversations with government officials and the Coalition to Save Canyonville, residents and officials raised the following health-related concerns:

1. Will children get cancer from their exposure to trichloroethene in private well water?

2. Are the birth defects and learning disabilities in Canyonville caused by chemicals from the site?

3. Should children play in yards along Canyon Road and in Canyon Wash?

4. Do residents need to move because of contamination in Horseshoe Canyon?

In addition, the Canyon County Health Department brought to our attention the fact that infant mortality rates for their county were elevated, and asked us to evaluate those rates. Although residents raised several questions about property values and liability issues, ATSDR is unable to address those questions in the context of the health assessment process because its focus is to evaluate the public health implications of the site.

public and for the scientific community. Thus, the language used to specify community health concerns should not indicate ATSDR's concurrence with or support of any allegations, before ATSDR has reviewed available information; rather, the assessor should use objective language to report community health concerns. Example 8. illustrates how to express those concerns. Finally, each concern in this section must be answered in the Community Concerns Evaluation subsection of the Public Health Implications section.

9.4. ENVIRONMENTAL CONTAMINATION AND OTHER HAZARDS

This section contains the foundation of the health assessment and, therefore, should be composed carefully so that the significant hazards of concern are clearly and concisely presented. In this section, the assessor describes (1) the contaminants that might pose a threat to public health and (2) physical hazards at the site.

For most sites, contamination may be clearly and conveniently designated as either "on site" or "off site." When possible, the designation of on site and off site should follow the definition in the site Remedial Investigation (RI). In some cases, this approach is inappropriate and some other definition may be needed. Further, for large sites with several components, or operable units, defining on site and off site may also be inappropriate, and other designations are necessary to clearly define the contamination. Cases in which the contamination is not discussed in the conventional on-site and off-site terms (as used in the RI), require that **all** such designations be clearly defined and delineated with the aid of appropriate location maps.

Regardless of the designation used, each subsection of the Environmental

Contamination and Other Hazards section should contain both text and tabular information. The text should identify, organize, and discuss contaminants of concern by medium, using subheadings such as surface soil, subsurface soil, sediment, groundwater, air, and soil gas. Environmental data should be presented in separate data tables labelled on site and off site. Tables should specify contaminants of concern, media, sampling date, reference source, and either maximum or both minimum and maximum concentrations. The assessor has the option of reporting the sampling date(s) in the narrative.

To determine whether a contaminant is a contaminant of concern based on noncancer end points, the maximum media concentration should be compared to an appropriate health assessment comparison value. Health assessors should use ATSDR's EMEGs, which are calculated from ATSDR's MRLs. If no EMEG is available, the assessor should use other health guidelines, such as EPA's RfD, to back-calculate a medium concentration. The assessor should also evaluate the potential carcinogenicity of contaminants. For carcinogens, comparison values based on a 10^{-6} cancer risk level for exposure to the contaminated media can be calculated from values such as EPA's Cancer Slope Factors. If the maximum medium concentration exceeds a comparison value, the contaminant should be selected for further evaluation. If a comparison value is not available, the contaminant should be selected.

The text should also discuss trends in the data (e.g., spatial distribution, hot spots, concentration changes over time, and contamination differences between media). Contaminant transport should not be discussed in this section.

9.4.1. On-Site Contamination

The On-Site Contamination subsection should define "on site" and discuss contamination within the site's boundaries. The text should discuss the following factors:

- sampling dates;
- spatial distribution of sampling locations;
- concentration changes over time (if relevant);
- medium-to-medium differences; and
- sample design and representativeness.

Sampling results in table form should follow the text. If sufficient information is available, the extent of contamination at the site can also be discussed. When a substance is identified as a contaminant of concern in one medium, its concentration should be provided for all media sampled. Tabulated values should reflect original data. Contaminant concentrations that have been included to document past exposures should be clearly identified in the text to distinguish them from current data. Footnotes can be used to clarify entries in the table or to denote data problems (e.g., quality assurance and quality control).

The assessor should discuss whether or not the sampling design and the number of samples adequately characterize on-site and off-site contamination. The text should identify deficiencies in sample design and number for each media and should discuss how these deficiencies affect the analyses, conclusions, and recommendations in the health assessment. This discussion can be presented either in this section or in the Pathways Analysis section. The text may also compare contaminant levels to local or regional values. Examples 9.A. and 9.B.

Example 9.A.

ON-SITE CONTAMINATION

EPA collected the data presented in this subsection during its site assessment in June 1987.

Waste Material

In two separate grab samples of sludge from the waste pit, EPA found high levels of trichloroethene (TCE), para chlorobenzenesulfonic acid (P-CBSA), mercury, and vinyl chloride (Table 1). Grab samples are usually taken from the surface of a medium without required depth or quantity. The sludge also contained other metals such as chromium, lead, and arsenic, but their concentrations were similar to soil concentrations upgradient of the site and to background soil concentrations for the western United States (5). Sludge levels of chromium and arsenic did not exceed our comparison values for health assessments. EPA did not analyze sludge samples or any other on-site or off-site media, except for off-site soil gas, for hydrogen sulfide. Because of the buried drums and the difficulty in collecting samples from the waste pit, EPA did not collect core samples in the pit. Without these data, we do not know whether or not the sampling data from the grab samples represent the valid range of concentrations in the waste pit. In addition, no information exists on the contaminants remaining in the waste pit after EPA's removal action.

Table 1. Range of Contaminant Concentrations in On-site Waste Materials

Contaminant	Concentration Range—ppm	Date	Reference	Comparison Value	
				ppm	Source
Mercury	1,500-4,000	6-87	(2)	15	RfD*
Chromium	23-65	6-87	(2)	200	EMEG
Arsenic	3-17	6-87	(2)	50	EMEG
Lead	15-44	6-87	(2)	none	none
TCE	200-150,000	6-87	(2)	127	CSF**
Vinyl Chloride	3-1,000	6-87	(2)	50	EMEG
P-CBSA	1,000-40,000	6-87	(2)	none	none

* EPA Reference Dose converted to a media-specific comparison value

** EPA Cancer Slope Factor converted to a media-specific 10^{-6} cancer risk level

Example 9.B.

SOIL

EPA collected five soil samples on the site using a 12-inch core. Because each core sample represented a mixed sample, ATSDR could not separate data on surface soil (less than or equal to 3 inches deep) from data on subsurface soil (more than 3 inches deep). Table 2 reports the contaminants and concentration range. While analyses of these on-site soil samples did not detect organic contaminants, such as TCE, vinyl chloride, and P-CBSA, the samples did contain several heavy metals: mercury, chromium, lead, and arsenic. The concentrations of chromium, lead, and arsenic are similar to concentrations upgradient of the site, and to background concentrations found in other soil from the western United States(5).

Five soil samples do not adequately characterize the existing contamination on this 69-acre site. Because no surface soil data are available for the site, a significant data gap exists in assessing the site's public health implications.

Table 2. Range of contaminant concentrations in on-site soil

Contaminant	Range of Levels—— ppm	Depth—— inches	Date	Reference	Comparison Value	
					ppm	Source
Mercury	625-900	0-12	6-87	(2)	15	RfD*
Chromium	32-68	0-12	6-87	(2)	200	EMEG
Arsenic	4.1-7.8	0-12	6-87	(2)	50	EMEG
Lead	8.7-23.6	0-12	6-87	(2)	none	none

* EPA Reference Dose (RfD) converted to a media-specific comparison value

provide sample presentations of text and tables.

9.4.2. Off-Site Contamination

Off-site contaminants of concern should be presented using the format for on-site contaminants. All contaminants identified in the on-site subsection should be included in the off-site subsection. On-site contaminants that have not been detected or reported off site should be identified. Example 10. illustrates a discussion of off-site contamination.

9.4.3. Quality Assurance and Quality Control

QA/QC information should be available through EPA's QA/QC summary and may be part of the RI/FS. QA/QC information should be used to determine whether the analytical techniques are adequate, the environmental data are representative, and the environmental media sampled are relevant. EPA has agreed to provide ATSDR with Data Review Summaries {Office of Solid Waste and Emergency Response (OSWER) Directive 9285.4-02}. If these summaries are not provided, they should be requested by contacting ATSDR's regional representative and having him/her request the QA/QC summary from EPA or, if necessary, from the

Example 10.

OFF-SITE CONTAMINATION

EPA obtained off-site data for soil, monitoring wells, private wells, surface water, sediment, and soil gas during its site assessment in June 1987. The CDHS provided past and current monitoring data (1985-1990) for the Canyon Spring Water Supply and the Canyonville Water Supply. From CDNR, we obtained the 1985 data for Santa River sediment and for fish.

Surface Soil

EPA collected five off-site surface soil samples in Horseshoe Canyon at a depth of 0-1 inch, which represents surface soil by ATSDR's definition. EPA obtained five samples from undeveloped land zoned for single-family residences. One of the five samples lies within the corporate limits of Canyonville, 500 feet northwest of Tots Nursery. While analyses of these off-site soil samples did not detect organic contaminants, such as TCE, vinyl chloride, or P-CBSA, the samples did contain several heavy metals: mercury, chromium, lead, and arsenic. The concentrations of chromium, lead, and arsenic in the downhill surface soil are similar to both the uphill concentrations and the background concentrations for the western United States (5). Spatial analyses indicate that mercury concentrations in off-site surface soils decrease from 130 to 40 ppm in the canyon south of Canyonville and to 20 ppm in Canyonville as one proceeds downhill from the site, thus suggesting the site is the possible source of contamination.

Five surface soil samples are inadequate for characterizing the existing surface contamination in Horseshoe Canyon. Because no surface soil data exist for residential yards, a significant data gap exists, and the health assessor is unable to assess the site's public health implications fully.

Table 3. Range of contaminant concentrations in off-site surface soils*

Contaminant	Concentration Range—ppm	Depth—inches	Date	Reference	Comparison Value	
					ppm	Source
Mercury	0.05-130	0-1	6-87	(2)	15	RfD**
Chromium	38-67	0-1	6-87	(2)	200	EMEG
Arsenic	3.2-6.7	0-1	6-87	(2)	50	EMEG
Lead	15-38	0-1	6-87	(2)	EMEG	NONE

* Samples collected outside of the creek's and wash's banks

** EPA Reference Dose converted to a media-specific comparison value

Example 11.

> We obtained the Quality Assurance/Quality Control (QA/QC) summary from EPA and reviewed the quality of their field and laboratory data. With the exception of one monitoring well (WM-6), which was not included in the health assessment, the QA/QC summary stated that field data and sampling quality during the EPA site assessment were satisfactory. Because of subsurface obstructions while installing WM-6, EPA did not complete construction of the well. No analytical problems were noted in the QA/QC summary except for acetone contamination from the laboratory in one groundwater sample and P-CBSA contamination in one monitoring well sample. The P-CBSA concentration in WM-2 is an estimated value because it exceeds the upper range of quantification. Because P-CBSA is found in other source media and in off-site groundwater, this one estimated value for P-CBSA and the other data quality problems discussed previously should not affect our analyses and conclusions about this site.

contractor. Every effort should be made to obtain QA/QC information. Absence of QA/QC information should be explicitly noted in this subsection of the health assessment. When QA/QC information cannot be obtained or does not exist, the health assessor needs to discuss the uncertainty and the limitations of the data and the effect the uncertainties and limitations have on the analyses, conclusions, and recommendations. The assessor may use or expand on the following qualifier:

> "In preparing this health assessment, ATSDR relies on the information provided in the referenced documents. We assume that adequate quality assurance and quality control measures were followed regarding chain of custody, laboratory procedures, and data reporting. The analyses, conclusions, and recommendations in this health assessment are valid only if the referenced documents are complete and reliable."

Example 11. illustrates a QA/QC discussion.

9.4.4. Physical and Other Hazards

In some instances, hazards other than those posed by chemical contamination may be present, including physical hazards (e.g., holes, lagoons, open steel tanks, abandoned materials, and equipment), special hazards such as the threat of fire or explosion, or hazards endemic to the site area. This subsection of the

Example 12.

> During our site visit, we observed evidence that bikers and garbage dumpers have access to the site. Because persons could fall in the waste pit, it is a physical hazard.

report should contain a brief description of such hazards and the potential health threat they represent. This description should also mention populations at risk from these hazards. Example 12. provides a brief description of physical hazards.

9.5. PATHWAYS ANALYSES

In the Pathways Analyses section, the health assessor will evaluate exposure pathways at the site. The introduction should provide a definition of the five elements in an exposure pathway:

1. source of contamination,

2. environmental media,

3. point of exposure,

4. route of human exposure, and

5. receptor population.

This section also should introduce the reader to the concept of completed and potential exposure pathways and should explain the difference between them. Completed exposure pathways exist when the five elements of a pathway link the contaminant source to an exposed population. Potential exposure pathways exist when information on one or more of the five elements is missing. Generally, the text should discuss completed exposure pathways in more detail than potential exposure pathways unless no completed pathways exist.

The assessor should divide the analyses into two subsections: completed exposure pathways and potential exposure pathways. Within each subsection, the text should identify each pathway with a subheading. The assessor should discuss each exposure pathway by explaining how contaminants migrate from the source to the exposed population. For example, to discuss exposure from drinking contaminated well water, the text should explain how contaminants migrated in groundwater from the source to private wells north of the site. The text should then explain that residents in the area with contaminated wells are exposed when they drink their well water. The text could also discuss other routes of exposure that are closely associated with this pathway, such as inhalation and skin exposure from bathing. The assessor could choose to discuss the pathways separately.

Completed and potential pathways should be presented in tables. Either the text or the table should indicate whether each pathway occurred in the past, is currently occurring, or will occur in the future. Example 13. illustrates how to present completed exposure pathways in a table.

An alternative to this approach would be to discuss the environmental component (e.g., source of contamination, environmental media, and point of exposure) of all pathways in one subsection. The remaining human exposure component (i.e., route of human exposure and exposed population) of these pathways would be discussed in a second subsection. With this approach, the assessor must ensure that the reader understands how the human exposure component of each pathway is linked to the appropriate environmental component of the exposure pathway. Because of the complexities involved in maintaining this link after splitting the exposure pathway into the two components, the health assessor is encouraged to use the first approach.

9.5.1. Completed Exposure Pathways

Completed exposure pathways exist when the five elements of a pathway link the contaminant source to a receptor population. In some cases, however, the original source of a completed exposure pathway may remain unknown. For example, it may be unclear whether the source of lead in tap water from a municipal water system comes from (1) indoor plumbing, (2) the distribution system, or (3) from groundwater. Nevertheless, the pathway is still considered complete.

To describe how a pathway leads to exposure, the text should explain how contaminants migrate from the source to the point of exposure. The text should also indicate how transport and transformation mechanisms affect movement of contaminants. When using technical language, the text needs either to define these terms or to use generally recognized words. The text should be limited to discussion of mechanisms that lead to human exposure. The discussion should be concise and specific and cover the following topics:

- contaminated media on and off the site (e.g., air, soil, groundwater, surface water, edible plants and animals, and wastes),

- the extent of contamination,

- relationships about transfer between contaminated media, and

Example 13.

PATHWAY NAME	EXPOSURE PATHWAY ELEMENTS					TIME
	SOURCE	ENVIRON- MENTAL MEDIUM	POINT OF EXPOSURE	ROUTE OF EXPOSURE	EXPOSED POPULA- TION	
Surface Soil	Canyon Waste Dump (CWD)	Surface Soil	Residential Yards and Playgrounds	Ingestion	Horseshoe Canyon Residents; Playground Users	Past Present Future
Sediment	CWD	Sediment	Santa River, Canyon Wash, and Canyon Creek	Ingestion	Children & Horseshoe Canyon Residents	Past Present Future
Fish	CWD	Fish	Residences	Ingestion	Santa River fish eaters	Past Present Future
Private Well	CWD	Groundwater (Private Well)	Residences, tap	Ingestion, Inhalation, Skin Contact	Residents Along Canyon Road South of Canyonville	Past Present Future
Public Spring	CWD	Groundwater (Public Well)	Residences, tap	Ingestion, Inhalation, Skin Contact	Canyon Springs Subdivision Residents	Present Future
Public Supply	Canyonville Water Supply	Municipal Water	Residences & Businesses in Canyonville, tap	Ingestion	Users of Canyonville Water Supply	Past Present Future
Ambient Air	CWD	Air	Nearby Yards & Buildings	Inhalation	Residents of Canyon Spring Subdivision & Canyon Road	Past Present Future

- possible and future contaminant migration.

Discussions of environmental fate and transport should not include all known geologic, topographic, hydrologic, climatologic, and other environmental information. Rather, the discussion should provide only information necessary for the reader to understand how contaminants migrate. When discussing the human exposure component, the text should indicate to the reader how human exposure occurs at the point of exposure. This explanation should include the routes of human exposure:

- ingestion—groundwater, surface water, soil, dust, food;

- inhalation—dust, vapor, gases;

- dermal absorption—contact with air, soil, dust, water, contaminated materials, or exposed wastes.

The discussion should include facts or estimates about duration and frequency of exposure and should specify the chemicals or chemical classes associated with each exposure pathway. Most important, it should also specify the location and the size of the population(s) exposed in the past and the present, and provide the same information on populations likely to be exposed in the future. The goal of the discussion is to show clearly how a particular population becomes exposed to specific contaminants.

The text should also indicate the significance of each pathway (i.e., how likely a pathway is to occur). The assessor may eliminate suspected exposure pathways that are unlikely to occur or that have only a remote possibility of occurring. The text, however, should briefly discuss pathways that are eliminated but which may be important to the public. The assessor will have to use his or her judgment to define unlikely, remote, and important. Because exposure pathways that are eliminated will not be discussed further in the health assessment, the assessor also should explain which chemicals have been eliminated. For exposure pathways that are eliminated but which are of community

health concern, the assessor should explain why they have been eliminated, however unlikely they may be as exposure pathways.

Discussion of environmental transport should be qualitative. The text may include a discussion of the physical and chemical properties of contaminants and media; it may also include quantitative assessments of environmental transport.

In addition to the exposure pathway, the assessor must also evaluate (either in this section or in the Environmental Contamination and Other Hazards section), whether or not the available data and information are sufficient to adequately characterize the five elements of the exposure pathway. The assessor should pay special attention to environmental conditions. Health assessors should determine if:

- sample location and number are representative of on-site and off-site contamination,

- analytic techniques are appropriate for the contaminants,

- environmental data are sufficient and relevant, and

- media sampled are appropriate.

The text should explain deficiencies in these areas and indicate media that have not been sampled. This discussion should concisely indicate how these deficiencies affect the analyses, conclusions, and recommendations. A more elaborate discussion is acceptable when the assessor believes that data gaps prevent conclusive discussion of public health hazards. In some cases, an appendix can be included when additional data needs are substantial. Example 14. provides a sample discussion of completed exposure pathways. Example 15. shows the presentation of the number of exposed and potentially exposed persons for each completed and potential pathway.

9.5.2. Potential Exposure Pathways

Potential exposure pathways exist when information or data on one or more elements

Example 14.

A. COMPLETED EXPOSURE PATHWAYS

Surface Soil Pathways

Past, current, and future exposure pathways may result from contamination of surface soils at several points of exposure: residential yards, undeveloped areas, playgrounds, and the site. While several points of exposure could occur in the present and in the future, we believe that only residential yards and playgrounds represent likely points of exposure (Tables 14 and 16). This site has several completed exposure pathways associated with surface soil. For convenience, we discuss all of these pathways under this subheading.

Contamination of the points of exposure has probably occurred because of several environmental transport mechanisms in Horseshoe Canyon. These mechanisms include transport of site contaminants by wind, surface runoff, flooding, excavation and construction, and fugitive dust from vehicular traffic. For example, heavy rains in 1974 and 1975 washed contaminants from the site's waste pit and contaminated surface soils into Canyon Wash. These storms and a 1990 storm flooded residential yards in Horseshoe Canyon, allowing contaminants from the site to settle onto the yards. EPA believes that these transport mechanisms are responsible for past and current releases of site contaminants from waste sources at the Canyon Waste Dump and for widespread mercury contamination of surface soils and other media (discussed below) in Horseshoe Canyon. Soil data indicate that mercury contamination has occurred in undeveloped and developed residential areas throughout Horseshoe Canyon.

For residential yards and facilities with playgrounds, such as Canyonville Elementary, soil ingestion is an important route of exposure, particularly for children less than 6 years old[4]. Soil ingestion is greater in young children because of their more frequent hand/mouth activity.

Surface soil mercury levels surrounding the site range from 40-130 ppm in residentially zoned areas south of Canyonville. In that area, 400 persons in Canyon Spring Subdivision are likely to be exposed to soil mercury levels from 40-90 ppm, and 200 persons along Canyon Road are likely to be exposed to soil mercury levels of 20-130 ppm. This conclusion about exposure is based on a limited number of surface soil samples (0-3 inches) in the Canyon and on a limited number of 0- to 12-inch deep soil samples from residential yards. Because EPA mixed the 0- to 12-inch deep core samples and included soil deeper than 3 inches in the samples, those samples probably underestimate the actual concentrations that exist near the soil surface, where exposure is more likely to occur. ATSDR requires surface soil data from residential yards to determine the degree to which exposure is occurring.

Because we have soil data on only three yards in Canyonville, we cannot assume that the results of those data are representative of the entire community of Canyonville. However, residents of the three yards in Canyonville are likely to be exposed to a mercury level of 5 ppm. Because the surface soil sample from the undeveloped residential property contained a mercury level of 20 ppm, soils at this location represent a future exposure pathway to residents through ingestion if the property is developed.

Example 15. Estimated Population for Exposure Pathways

Exposed Populations and *Potentially Exposed Populations* that are affected by a Completed or *Potential Exposure Pathway**

Location			Estimated persons	Mercury	TCE	P-CBSA	Lead**
Canyon Road Residents			200	Surface Soil, Sediment	Private Well, Ambient Air	Private Well	Not Known
Tots Nursery Children			*125*	*Sand, Surface Soil*	*Not Known*	*Not Known*	*Not Known*
Children at Canyonville Elementary			249	Surface Soil, Sediment	Not Known	Not Known	Not Known
Santa River Fish Eaters	Canyon Road South of Canyonville	Past	75	Fish	Concurrent Exposure	Concurrent Exposure	Not Known
		Present	0				
	Other	Past	Not Known				
		Present	250				
Residents of Canyon Spring Subdivision			400	Surface Soil Sediment	Public Spring, Air	Not Exposed	Not Known
Persons Along Canyon Creek Downstream of Canyon Wash	Residents		10	Soil, Sediment	Not Exposed	Not Exposed	Not Known
	Playing Children		Not Known	Soil Sediment	Not Known	Not Known	Not Known
Canyonville Households Having Taps Sampled			7	*Canyonville Soil*	Not Exposed	Not Exposed	Not Known
Canyonville Residents			*36,029*	*Canyonville Surface Soil*	*Not Exposed*	*Not Exposed*	*Public Water Supply*
Site Workers			*Not Known*	*Worker-Waste Material*	*Worker-Waste Material*	*Worker-Waste Material*	*Not Known*

*Potential exposure pathways are presented *in italics*

** Unrelated to site

in a pathway are missing. Modeled data cannot be substituted for actual data; therefore, when only modeled data are available, the pathway is considered potential.

The assessor should present and discuss potential exposure pathways in the same format as completed exposure pathways. However, the assessor may choose to limit the discussion until more information or data become available.

9.5.3. Alternative Approach for Pathways Analyses

When choosing the alternative approach (the environmental component of each pathway is discussed separately from the human exposure component of a pathway), the assessor will use two subsections:

A. Environmental Component (Fate and Transport)

B. Human Exposure Component.

The assessor still must identify the completed and potential exposure pathways. As in the preferred approach, the Environmental Component subsection should discuss the source of contamination, media and transport, and point of exposure. The Human Exposure Component subsection should identify the route of human exposure and the receptor population. The human exposure component of each pathway should be presented in the same order as the environmental component, and the link between the two components of each pathway should be clear.

9.6. PUBLIC HEALTH IMPLICATIONS

In the Public Health Implications section, the health assessor will discuss the health effects of site contaminants, evaluate health outcome data, and address all questions raised by the community. Accordingly, this section has three subsections:

* Toxicologic Evaluation,

* Health Outcome Data Evaluation,

* Community Health Concerns Evaluation.

When using medical, toxicologic, and epidemiologic terms, the assessor should define the terms or use generally recognized words so that the reader will understand.

The reader must be able to understand how information in each section of the health assessment relates to the public health discussion. The assessor should therefore incorporate pertinent information from previous sections to support the evaluations in the Public Health Implications section, thus making the health assessment site-specific. For example, facts about the site's history, operation, and remediation, as well as information about demographics, land use, and natural resource use, could be important in evaluating health issues arising from contaminant exposure. Likewise, the assessor must link the pathways analyses with the public health implications discussion. For example, when reading about effects of lead toxicity, the reader needs to know who the exposed population is and if this population may experience those health effects.

9.6.1. Toxicologic Evaluation

The assessor should organize the Toxicologic Evaluation subsection by contaminant or by route of human exposure rather than by media. This subsection should build upon the completed and potential exposure pathways identified in the Pathways Analyses by linking specific contaminant exposure in certain populations with the discussion of health effects. Contaminants associated with completed exposure pathways should receive more attention than contaminants in potential exposure pathways. However, the assessor may emphasize health discussions of contaminants in potential exposure pathways if the pathway is likely to become a completed pathway. When only potential exposure pathways exist, the discussion of health effects from potential exposure may be expanded. The assessor should use the exposed population in each exposure pathway to determine the appropriate parameters for estimating dose. For example, when the point of exposure for a completed exposure pathway is residential yards, the assessor must use soil ingestion rates for children as well as adults.

Using appropriate parameters, the assessor should estimate an exposed dose for each contaminant in a completed exposure pathway. If information is available, the assessor should also estimate the exposed dose from potential exposure pathways. When the same receptor population is exposed to a contaminant through multiple pathways, the assessor should consider the health effects from the total exposed dose. The assessor should qualitatively describe and compare the exposed dose with ATSDR's acute, intermediate, and chronic Minimal Risk Levels (MRLs). When MRLs are unavailable, other health-based guidelines, such as EPA's RfD, are acceptable. The text may include these health-based guidelines and the specific value in a table.

Example 16. illustrates these guidelines using a table.

Using health-based guidelines, the text should describe health effects likely to occur from the exposure (and the estimated dose) that occurred, is occurring, or will occur. This evaluation should be based on:

- population affected (e.g., on-site workers, residents of a specific area, and trespassers),

- routes of human exposure (e.g., inhalation, ingestion, skin contact),

- chemical involved (individually or grouped according to chemical class or toxic effect),

- acute health effects, if likely (for short-term exposure periods to each chemical or group), and

- chronic health effects, if likely (for short- and long-term exposure periods to each chemical or group).

Additionally, sensitive subpopulations, the effects of chemical interactions, and personal actions that can reduce exposure should be identified when appropriate. ATSDR's

Example 16.

Comparison of Estimated Exposed Dose to Health Guidelines for Mercury and Methylmercury				
Contaminant	Exposure Pathway	Health Guideline for Ingestion - mg/kg/day		
		Value	Source	Exceeded by Estimated Exposure Dose
Methylmercury	Fish	0.00004	Acute MRL	yes
Methylmercury	Fish	0.00002	Intermediate MRL	yes
Mercury	Soil and Sediment	0.2	Acute MRL	no
Mercury	Soil and Sediment	0.0008	Intermediate MRL	yes

Toxicological Profiles serve as useful sources of environmental and health information. The profiles and other sources should be cited in the discussion.

The text should include discussions explaining health effects according to biological plausibility:

- route of exposure,

- duration of exposure,

- exposed dose,

- dose-response relationships, and

- variation in pharmacokinetics.

The text should clearly state when health discussions are based on toxicologic information that comes from another route of exposure. Additionally, results from animal and human studies, such as high dose in animals or unknown dose in humans, should be compared and contrasted with site-specific exposure conditions to determine which health effects are likely to occur.

When persons are exposed to carcinogens from the site, or when the potential for exposure exists, the text should include a qualitative discussion on whether or not cancer is likely to occur from that exposure. Many factors, such as dose, population sensitivity, duration, and frequency of exposure affect whether cancer is possible. The assessor may include those factors in the discussion.

When toxicologic, environmental, epidemiologic, or other information is missing, these data gaps are important to the discussion of whether or not health effects are likely to occur. When information is insufficient to reach health conclusions, the text should provide a qualified discussion of health effects for the exposed population and the exposure pathway involved.

Example 17. illustrates how to present many, but not all, of the concepts discussed. The assessor should always tailor the health discussion to the specific exposure pathways and conditions identified at the site.

9.6.2. Health Outcome Data Evaluation

The health assessor should evaluate the local, state, and national databases previously described in the Background section that are relevant to the site. The Health Outcome Data Evaluation subsection reports the results of the epidemiologic analysis of these databases. The text should be organized according to the name of the database. For each database, the text should include the following information:

- the reasons for the evaluation (e.g., the outcome is plausible, the community requested it),

- database characteristics (e.g., available years of data, smallest geographic unit in database),

- relationship of exposed population to smallest geographic unit,

- characteristics of the comparison population,

- methods of analysis (e.g., SMR),

- results of the analysis,

- limitations of the results (e.g., data reliability, effect of geographical unit on interpretation), and

- effect of results on public health implications.

As discussed in Chapter 2, health outcome data should be as site-specific as possible. A common problem in using secondary health outcome data is that the exposed population at the site will be part of a greater population covered by the database. Thus, morbidity and mortality rates calculated from health outcome data usually will not represent the true rates for the exposed population. The text, therefore, needs to qualify the interpretation of rates.

Nevertheless, reporting the rates of relevant health outcomes will answer, to some degree, community concerns about specific adverse health effects. Rarely, if at all, will the assessor be able to identify causal relationships between adverse health effects in these databases and

Example 17. Public Health Implications

PUBLIC HEALTH IMPLICATIONS

A. TOXICOLOGIC IMPLICATIONS

Mercury

Some residents inside and outside Horseshoe Canyon were exposed to mercury and methylmercury through several completed exposure pathways. Before the state health advisory in 1985, exposure to methylmercury occurred in adults and children who ate contaminated fish. From the country's well survey, we know that residents along Canyon Road typically ate zero to three fish meals each month from the Santa River. To estimate acute methylmercury exposure, we assumed that children, women, and men ate 4, 8, and 13 ounces of fish per meal, respectively.

More than 80% of the mercury in freshwater fish is methylmercury. We will assume that the total mercury measurement for Santa fish consists of methylmercury. The estimated amount of methylmercury ingested by children and adults who eat the most highly contaminated fish in the Santa River (for example, trout), exceeds ATSDR's ingestion MRL for acute and intermediate exposure to methylmercury. Therfore, persons who have eaten and are eating Santa River fish may experience harmful health effects.

The brain, the kidneys, and fetuses are most sensitive to methylmercury exposure. Neurologic symptoms in adults and children with brain damage include 1) prickling, tingling sensations in the arms and legs, 2) loss of sensation in the arms and legs, 3) tunnel vision, 4) slurred speech, 5) incoordination, 6) irritability, 7) memory and hearing loss, and 8) difficulty sleeping (ATSDR Toxicological Profile for Mercury). While kidney data in humans are limited, animal studies have shown that methylmercury can damage the kidney. Kidney effects include increased urine output and elevated urine levels of albumin, a blood protein which is normally not present in urine. Human studies on pregnant women have shown that methylmercury exposure causes neurologic affects in offspring such as mental retardation, incoordination, and inability to move. Milder health effects include delayed neurologic development and slower muscular movements. Futhermore, animal studies of methylmercury exposure during pregnancy show problems with behavioral maturation and learning ability in offspring. Those health effects could occur in children of women who have eaten or are eating contaminated fish from the Santa River.

Children and adults in Horseshoe Canyon who ate mercury-contaminated fish before the state health advisory was issued were also exposed to mercury through the ingestion of mercury-contaminated soil and sediment. The sum of those exposures before the health advisory may increase the frequency and severity of health effects in those children and adults.

Some residents outside Horseshoe Canyon are continuing to eat mercury-contaminated fish from the Santa River. We suspect their fish consumption patterns to be similar to historical patterns reported by residents in Horseshoe Canyon, though we cannot be certain. Their health effects could be similar to the ones discussed earlier, and, without intervention, their exposure will continue.

Residents who live adjacent to Canyon Wash and Canyon Creek downstream of Canyon Wash and who ingest contaminated sediment and soil have been and continue to be exposed to mercury. The estimated mercury ingested by children and adults from soil ingestion does not exceed ATSDR's acute MRL for ingestion. Health effects, therefore, are unlikely to occur if soil ingestion is infrequent. However, for young children, particularly those under six years old, who play along Canyon Wash every day, their mercury exposure could exceed ATSDR's intermediate MRL. This exposure could damage the kidney and lead to neurologic effects. Because of lower soil ingestion, adults and older children are unlikely to experience health effects from contaminated soil and sediment along Canyon Wash and Canyon Creek. Their intake of mercury from soil does not exceed ATSDR's intermediate MRL.

exposure to site contaminants. In such cases, the assessor should decide if the evaluation of epidemiologic, toxicologic, and pathways analyses can be used to support the need for follow-up health studies. Example 18. provides a sample discussion of health outcome data evaluation.

9.6.3. Community Health Concerns Evaluation

The health assessor should use the pathways analyses, the toxicologic evaluation, and the health outcome data evaluation to answer all site-related community health concerns. In responding to community health concerns, the text should restate the concern. For concerns that are plausible (i.e., are possible because of contaminant exposure from the site), the assessor should answer the question using information presented in the Toxicologic Evaluation or the Health Outcome Data Evaluation subsections. The assessor should refer the reader to those subsections for further details. For concerns that are not plausible, the assessor should answer the question using

toxicologic information about the contaminant and, if available, health outcome data. For example, when a resident raises a question about liver cancer in the community, the assessor should use a cancer registry, if available, to determine if liver cancer rates are elevated. In addition, the assessor should discuss whether persons are exposed to a contaminant that causes liver cancer and whether or not that exposure could cause liver cancer. Example 19. illustrates an acceptable approach for responding to community health concerns. All questions raised by the community must be addressed in this subsection. However, questions about economic and liability issues (e.g., property values, installation fees, legal questions) should be handled by stating that ATSDR's health assessment addresses only site-related health issues.

Health questions may arise that are unrelated to the site. Those questions should be addressed in the Community Health Concerns Evaluation subsection. When such questions arise, the assessor needs to judge the type and

Example 18. Health Outcome Data Evaluation

To address the community's concern about the high number of birth defects in Greater Canyonville, information from Calizona's Birth Defects Registry for 1986, 1987, and 1988 was evaluated. In 1986 and 1987, fewer children in Greater Canyonville were born with birth defects than were children in the rest of the county and state. Using the same comparison populations (county and state), data from 1988 showed a greater incidence of birth defects than expected in Greater Canyonville (Standard Incidence Ratio = 1.44).

Based on the toxicologic data reviewed, exposure to mercury, lead, and TCE at levels present at the Canyonville Waste Dump site is unlikely to cause adverse outcomes that are typically classified as physical birth defects, such as cleft palate and spina bifida. Several limitations are inherent in this analysis. The database's short period of existence makes the conclusions unreliable. In addition, toxic effects from exposure to lead, TCE, and mercury may have occurred *in utero,* leading to adverse neurological effects in infancy. Those adverse health outcomes are not routinely diagnosed at birth and would therefore not be captured in this birth defects registry.

A preliminary review of the birth defects registry indicates there is no clear indication of elevated rates of birth defects in Greater Canyonville during the period 1986-1988. Moreover, physical birth defects are not likely to occur based on the pathway analysis and toxicologic information.

Example 19.

C. COMMUNITY HEALTH CONCERNS EVALUATION

We have addressed each of the community concerns about health as follows:

1. Will children get cancer from their exposure to trichloroethene (TCE) in private well water?

Although residents near the site along Canyon Road use and have used TCE-contaminated water, no studies have shown the TCE causes cancer in humans. However, because animal studies have shown that TCE causes cancer, we assume that TCE could be a human carcinogen. Based on our conservative estimates, children and adults who drank and bathed in water containing TCE levels present in the two private wells on Canyon Road could, over a lifetime, be at a low increased risk of developing cancer. While the number of extra cancers above the background cancer number theoretically may have increased slightly, TCE-induced cancer in residents of those two households is unlikely to occur. Until more data are collected on other private wells on Canyon Road, we do not know the extent of the risk to other residents. ATSDR has a follow-up registry of persons exposed to TCE by ingestion, which documents health effects. ATSDR will consider including TCE-exposed residents in this registry.

2. Are the birth defects and learning disabilities in Canyonville caused by chemicals from the site?

The contaminants at Canyon Waste Dump are not known to cause physical birth defects. We evaluated the Calizona Birth Defects Registry and found no increase in the rate of physical birth defects for greater Canyonville.

At this time, we cannot determine whether the higher percentage of learning disabilities at Canyonville Elementary is related to the site. The higher percentage in Canyonville Elementary students may have resulted from the school including gifted and emotionally disturbed children in its definition of learning disabilities. We have recommended further investigations of learning disabilities.

3. Should children play in yards along Canyon Road and in Canyon Wash?

Until further evaluation is complete, children should not play in Canyon Wash or Canyon Creek. When we have data for each residential yard, we can address this question more completely. We have asked EPA to collect surface soil samples from each yard that borders Canyon Wash and Canyon Creek.

In addition to these questions, residents raised several questions about property values and liability issues. ATSDR is unable to address those questions in the context of the health assessment because the Agency's focus is to evaluate the public health implications of the site. We suggest that residents discuss those issues with EPA and with their personal lawyers.

degree of response necessary for each question. For example, a question about magnetic radiation from nearby power lines that are not part of the site requires only that the assessor, when possible, direct the public to the appropriate authorities who can address the question. It does not require a health discussion of magnetic radiation. In contrast, off-site sampling may have detected lead in tap water that is not related to the site. In this case, the assessor may choose to discuss lead toxicity in the Toxicologic Evaluation subsection, and make recommendations to prevent exposure. The assessor may also recommend follow-up health activity, or refer the public to local health authorities.

- health effects from exposure to site contaminants,

- response to community health concerns,

- results of health outcome data evaluation, and

- the effect that missing or insufficient information has on analyses and conclusions.

All conclusions should be explicit and unambiguous, stating concisely the findings of the health assessment (see Example 20.). Every conclusion of the health assessment should have one or more recommendations associated with it.

9.7. CONCLUSIONS

The final section of the health assessment should address conclusions about the site and the health threat it poses. The first conclusion should be a statement about the site's level of public health hazard. The assessor should assign one of the five public health categories:

- urgent public health hazard,

- public health hazard,

- indeterminate public health hazard,

- no apparent public health hazard, or

- no public health hazard.

For the categories "urgent public health hazard" and "public health hazard," the text should identify the contaminant(s), the completed exposure pathway(s), the health effect(s), and the exposed population(s). This section should briefly summarize why the category was chosen. New information should not be discussed in this section.

The text should also summarize conclusions about the following issues:

9.8. RECOMMENDATIONS

9.8.1. Recommendations

ATSDR may recommend that actions be taken to protect public health. The assessor may make recommendations that:

- end or reduce exposure,

- characterize the site, or

- suggest follow-up health activity.

All recommendations should be numbered, start with an action verb, and follow parallel construction. In addition, recommendations should correlate with the conclusions in the Conclusions section.

Recommendations made because of health effects that result from exposure to site contaminants should identify actions that prevent or reduce exposure. Risk management recommendations, such as constructing a fence or installing a berm, should be avoided. Instead, recommendations should be made so that response options are not specified (e.g., restrict access to the site, prevent surface migration).

When environmental data and other information are insufficient to determine the public health hazard posed by the site, recommendations for data gathering are required. The assessor should identify the data needed, where the data should be collected, and who should receive the data.

Each recommendation should express the urgency with which or timeframe in which the data gathering should be accomplished. The urgency may be expressed directly (e.g., alternative water supplies should be provided immediately) or set within the time frame of other activity (e.g., during remediation). Recommendations that do not have a timeframe for completion may be interpreted as having low priority.

Example 20.

The Canyon Waste Dump site poses an urgent public health hazard because acute, intermediate, and chronic exposure to mercury from eating contaminated fish from the Santa River could cause adverse health effects. Damage could occur to kidneys, fetuses, and the brain.

Furthermore, a public health hazard exists for young children who lived or are living along Canyon Wash. Because of mercury exposure from ingesting soil and sediment, they may suffer adverse health effects such as kidney and neurologic damage. While lead in the Canyonville Public Water supply is unrelated to the site, it is a public health hazard. Lead-contaminated water can cause adverse health effects in infants and young children, resulting in decreased IQ scores, growth retardation, and hearing problems.

We have received many health questions about this site from residents and officials in Horseshoe Canyon. Those concerns are summarized and addressed in the Public Health Implications section.

We reviewed community-specific health outcome data for learning disabilities and found that learning disabilities are higher at Canyonville Elementary than at other schools in greater Canyonville. Even though learning disabilities can be caused by mercury and lead, we cannot link the school's higher rate of learning disabilities to the site without better epidemiologic information.

Data inadequacies include the following:

1. Environmental data for surface soil (less than or equal to 3 inches deep), sediment, and groundwater from monitoring wells and private wells do not adequately characterize the extent or amount of contamination that may exist on and off the site. See the Environmental Contamination and Other Physical Hazards section for details.

2. No data exist documenting mercury levels in (1) surface soils of residenctial yards, (2) playground surfaces such as sandbox sand at Tots Nursery, (3) indoor dust on hard surfaces, or (4) particles in ambient air.

The final recommendation in every health assessment should pertain to follow-up health actions. Three types of actions exist: environmental health education, health studies, and substance-specific applied research. Additional information is provided in Chapter 8.

Example 21. illustrates several acceptable recommendations.

Example 21.

RECOMMENDATIONS

Cease/Reduce Exposure Recommendations

1. Inform Horseshoe Canyon residents immediately that no children should play in (1) Canyon Creek downstream of the confluence with Canyon Wash, and (2) the Tots Nursery sandbox until authorities verify those areas are uncontaminated.

2. Provide uncontaminated water immediately for all domestic purposes for the two households with contaminated wells, as well as any other households found to have contaminated wells.

Site Characterization Recommendations

1. Obtain additional data for surface soil (less than or equal to 3 inches deep) and sediment to characterize adequately the extent and amount of site contamination that may exist on and off the site. In particular, perform the following:

   a. Collect additional samples of surface soils immediately for (1) each residential yard along Canyon Road and in Canyon Spring Subdivision that borders Canyon Wash, (2) each residential yard that borders Canyon Creek downstream of its confluence with Canyon Wash, (3) 50 residential yards in Canyonville, and (4) playground areas at Canyonville Elementary and Tots Nursery, including the Tots Nursery sandbox.

   b. Collect 10 upstream and 30 downstream fish samples in the Santa River and its tributaries and analyze edible fish portions for total mercury and methylmercury. Collect fish samples up to 10 miles downstream.

Health Activities Recommendation Panel (HARP) Recommendations

In accordance with the Comprehensive Environmental Response, Compensation, and Liability Act of 1980 as amended, ATSDR and the state have evaluated the Canyon Waste Dump Site for appropriate health follow-up activities. ATSDR is currently issuing a health advisory regarding consumption of fish from the Santa River. In addition, the panel offers the following recommendations:

1. Provide immediate community health education to the exposed populations in greater Canyonville about the possible health effects from site contaminants and interim measures to reduce exposures.

2. Educate health professionals in greater Canyonville about the health effects of the site contaminants.

3. Conduct a follow-up exposure study on past and current eaters of Santa River fish.

4. Consider additional follow-up activities if data become available that suggest human exposure is occurring or has occurred.

9.8.2. Public Health Actions

Based on the recommendations presented in the health assessment, the health assessor needs to identify actions that have been undertaken or that are planned. In addition, the health assessor should identify the agencies that have conducted, and that plan to conduct, specific actions. The assessor should (1) work through the Health Activities Recommendation Panel (HARP) to identify the actions that ATSDR will conduct; and (2) work with other federal, state, and local agencies to identify the actions that they will

Example 22.

Public Health Actions

Based on the recommendations made in the health assessment, the following public health actions have been or will be undertaken.

Actions Undertaken

1. The Division of Health Education, ATSDR, is providing environmental health education to health professionals in greater Canyonville. The division is also informing the community about adverse health effects that may occur as a result of eating mercury-contaminated fish from the Santa River. This program should be complete by mid-year 1992.

2. The Canyon County board of commissioners passed in January 1992 an ordinance prohibiting the installation of private wells in Canyon County. That ordinance will prevent exposure to the contaminated groundwater.

3. The Environmental Protection Agency (EPA) has collected surface soil samples (<3 inches) from residential and potential residential areas in Canyonville. Sample collection was completed by December 1991. After laboratory analysis and quality assurance/quality control are completed, ATSDR, the Calizona Department of Health Services, and the Canyon County Health Department will evaluate the results to determine the public health significance of surface soil contamination in residential areas.

Actions Planned

1. The Calizona Department of Natural Resources, in cooperation with the U.S. Fish and Wildlife Service, plans to collect fish samples from the Santa River. The results will help public health officials confirm the current degree and extent of fish contamination in the Santa River. The results also will be used by ATSDR as part of an exposure study it plans to conduct.

2. ATSDR, in cooperation with the Calizona Department of Health Services and the Canyon County Health Department, will collect blood and hair samples to determine whether residents who have been eating fish from the Santa River have had higher than expected exposure to methylmercury. The study will start in the Fall of 1992.

3. EPA has agreed to collect 50 random samples of tap water from residences connected to the Canyonville Water Supply. EPA has agreed to collect the samples by the end of 1992. Public health officials will use the data to determine whether further public health actions are needed to prevent lead exposure from drinking tap water.

---

conduct. The purpose of this approach is to delineate in the health assessment the site-specific health agenda that will mitigate or prevent adverse health effects in humans that could result from exposure to hazardous substances.

The assessor should identify these actions in a "Public Health Actions" section. Within that section, actions should be categorized as "actions undertaken" and "actions planned."

Regarding actions undertaken, the assessor should indicate actions or activities that have occurred or that are currently occurring because of recommendations in the health assessment. These actions may result from interactions between the health assessor and the respective agencies as he or she develops recommendations for the health assessment, or

they may result from the review process of the health assessment. For example, in response to the assessor identifying a data gap in soil contamination, EPA's remedial program manager may have collected additional surface soil samples (0-3 inches) so that ATSDR could more adequately evaluate the public health implications of soil ingestion in children. The assessor would identify that activity as an action that EPA undertook as a result of interaction with the health assessor. The assessor should identify (1) the action, (2) the agency or group that conducted the action, (3) the purpose of the action, and (4) the date the action occurred.

For actions planned, the same philosophy applies. Based on recommendations in the health assessment, the assessor should identify (1) the action, (2) the agency or group that will conduct the action, (3) the purpose of the

Example 23.

PREPARERS OF REPORT

Name
Title
Branch/Office/Division
Agency Name

Name
Title
Branch/Office/Division
Agency Name

ATSDR TECHNICAL PROJECT OFFICER

Name
Title
Branch
Division

ATSDR REGIONAL REPRESENTATIVES

Name
Title
Branch
Division

Example 24.

REFERENCES

1. Calizona Department of Health Services. Data sheets concerning contamination of Santa River Fish Data. November 1985.

2. EM Company. Site Assessment, Canyon Waste Dump, Canyon County, Calizona. June 1987.

3. Canyon County Health Department. Well survey of Horseshoe Canyon residents. September 4-6, 1990.

4. Environmental Protection Agency. Exposure Factors Handbook. Washington, DC: Environmental Protection Agency, Office of Health and Environmental Assessment, July 1989; EPA document no. 600/8-89/043.

5. Agency for Toxic Substances and Disease Registry. Draft Health Assessment Guidance Manual. Atlanta, Georgia: Agency for Toxic Substances and Disease Registry, February 1991; DHHS (PHS).

6. Agency for Toxic Substances and Disease Registry. Draft Toxicological Profile for Total Xylenes. Atlanta, Georgia: Agency for Toxic Substances and Disease Registry, December 1990; DHHS publication no. (PHS)TP-90-30.

7. Wallace LA. 1987. Total Exposure Assessment Methodology (TEAM) Study: Summary and Analysis: Volume I. Washington, DC: U.W. Environmental Protection Agency, Office of Acid Deposition, Environmental Monitoring and Quality Assurance, June 1987; EPA document no. 600/6087/002a.

action, and (4) if known, the timeframe for conducting the action. The actions planned will generally consist of two types: activities that HARP determines are appropriate and activities that other agencies agree to conduct. The Public Health Actions section should clearly identify the agencies involved in conducting the actions.

Example 22. illustrates some public health actions.

9.9. PREPARERS OF REPORT

Every health assessment should conclude with the signature blocks (Example 23.) of the health assessment team. This team includes (1) the preparers of the report, (2) the ATSDR regional representative, and, for state-prepared health assessments, (3) the

ATSDR Technical Project Officer. The signature block should also include the job title and the appropriate agency name.

9.10. REFERENCES

The Reference section (Example 24.) should list the documents reviewed and the sources of information for environmental data, community concerns, health outcome data, and toxicologic data. The reference style used comes from the Uniform Requirements for Manuscripts Submitted to Biomedical Journals.

References should appear in the text, tables, and legends using Arabic numerals within parentheses. References are numbered consecutively according to the order in which they appear. Complete citations for all references should be provided. Citations are numbered consecutively according to where they appear in the text. If references are not cited in the text, a list arranged alphabetically and entitled "Selected Bibliography" should be included at the end of the assessment.

9.11. APPENDICES

Material attached to the health assessment as appendices should be referenced in the text. A map showing site characteristics is necessary for the reader to visualize on-site and off-site features. Maps taken from other documents should be reworked to fit the discussion in the text. The assessor should ensure that features labelled on the map are discussed in the text.

After a health assessment has undergone public comment, but before the assessment becomes final, the health assessor's response to public comments should be added as an appendix to the health assessment.

Appendices should be used judiciously and only when special site-specific conditions warrant them. Each appendix should be identified in a subheading in this section (including appendix letter identifier); individual appendices should be labelled with corresponding titles.

Environmental Media Evaluation Guides (EMEGs) are media-specific comparison values that are used to select contaminants of concern at hazardous waste sites. The use of EMEG values in the health assessment process is described in Section 5.6. of this manual.

EMEGs have been calculated for chemicals for which ATSDR has developed Toxicological Profiles. These chemicals were selected because of their toxicity, frequency-of-occurrence at sites or facilities on the National Priorities List (NPL), and potential for human exposure to the substance. The ATSDR EMEGs will be periodically updated as the Division of Toxicology releases additional Toxicological Profiles and revises older ones.

EMEGs are derived from the Minimal Risk Levels (MRLs) presented in the ATSDR Toxicological Profiles. The Toxicological Profiles and the MRLs have undergone internal ATSDR review as well as external peer review by a panel of scientific experts. An MRL is defined as an estimate of daily human exposure to a chemical that is likely to be without an appreciable risk of deleterious effects (noncarcinogenic) over a specified duration of exposure. Thus, MRLs provide a measure of the toxicity of the chemical.

In addition to containing a toxicity component, the EMEGs also contain an exposure component that is based on the amount of contaminated water or soil that an individual ingests per day. Because water consumption and soil ingestion vary widely in different segments of the population, EMEG values are calculated for a range of exposures rather than for a single, arbitrary exposure value.

The range of the exposure parameters used for calculating EMEGs is discussed separately under each environmental medium. Exposure

to contaminated water or soil is usually greater in children than in adults because children typically ingest more water and soil per unit of body weight than adults. Therefore, at sites where both children and adults are present, EMEG values derived for children are usually used, because they represent the more highly exposed population. However, health assessors are encouraged to consider all available site-specific information when selecting the appropriate EMEG value to use for selecting contaminants of concern.

In the ATSDR Toxicological Profiles, MRLs are developed for acute, intermediate, and chronic exposure intervals. Acute exposures are defined as those of 14 days or less; intermediate exposures are those lasting more than 14 days but less than one year; and chronic exposures are those lasting one year or longer. For health assessment purposes, it is usually assumed that chronic exposures are possible; therefore, EMEGs should be derived from the chronic MRLs. However, when chronic exposures can be excluded, health assessors have the option of deriving EMEGs from an acute or intermediate MRL.

Estimating the health impact of exposure to chemical mixtures is of particular concern because hazardous waste sites often contain multiple chemical contaminants. There is a considerable body of scientific information documenting the occurrence of interactive effects from simultaneous exposure to two or more chemicals. Such interactions may be additive, antagonistic, or synergistic. However, for most chemical mixtures, information on toxic interactions is lacking. Furthermore, even though limited information for some chemical mixtures is available, no set of evaluation guides could account for the infinite array of chemicals in varying proportions that may be found at sites. Therefore, EMEGs are based on exposure to a single chemical and do not consider the effects of exposures to chemical mixtures.

One approach that is sometimes used to estimate the toxicity of chemical mixtures is to assume that the chemicals have an additive toxic effect. This approach may be particularly relevant if the chemicals in the mixture have a common mechanism of action or affect the same anatomical site. Further information on the use of additivity models to estimate the toxicity of chemical mixtures can be obtained from published references (1,2). The scientific basis for the application of these models to predict the toxicity of chemical mixtures is very limited. Therefore, the use of such models is left to the discretion and professional judgment of the health assessor.

In the derivation of EMEGs, it is assumed that exposure is occurring from a single medium. However, it is important to recognize that a person could be concurrently exposed to a chemical from several exposure pathways. For example, a contaminant in water might also be present in food, soil, air, etc., and exposure to those other media could increase the total dose. The relative contribution of a particular exposure pathway to the total dose could vary dramatically depending on site-specific circumstances. Because of site-to-site variability, it is not feasible to propose EMEGs that account for possible exposures from other media. Therefore, if exposure to a contaminant is occurring by multiple exposure pathways, the health assessor should add together the exposures from various pathways to determine the total body dose.

As discussed previously, MRLs are based on non-carcinogenic toxic effects of chemicals, including their developmental and reproductive toxicity. MRLs do not consider the potential genotoxic or carcinogenic effects of a chemical. Currently, the ATSDR Division of Toxicology uses a weight-of-evidence approach in its assessment of the carcinogenic hazard posed by substances that are the subject of Toxicological Profiles. As part of this assessment, the Division of Toxicology relies heavily on conclusions drawn by the National Toxicology Program for substances that have been evaluated in its Annual Report on Carcinogens (3). Additionally, conclusions drawn by the International Agency for Research on Cancer (IARC) and the U.S. Environmental Protection Agency (EPA) are considered. However, as part of its overall evaluation of hazard for all substances (including substances that have been previously evaluated by those three groups), ATSDR does evaluate and draw conclusions regarding the substance's oncogenic potential.

A.1. WATER EVALUATION GUIDES

Water EMEGs can be derived for potable water used in the home. Potable water refers to water used for drinking, cooking, and food preparation. Water used in the home for nonpotable purposes such as cleaning, washing, bathing, and showering is also discussed here. This discussion does not consider water used for irrigating crops, watering livestock, swimming, or other purposes.

For potable water exposures, an EMEG is derived from the following equation:

$$EMEGw = \frac{MRL \times BW}{IR}$$

where,

EMEGw	= water evaluation guide (mg/L)
MRL	= minimal risk level (mg/kg/day)
BW	= body weight (kg)
IR	= ingestion rate (L/day)

To derive the water EMEG for a chemical, use the chronic oral MRL from the ATSDR Toxicological Profile. Ideally, the MRL should be based on an experiment in which the chemical was administered in water. However, in the absence of such data, an MRL based on an experiment in which the chemical was administered by gavage or in food could be used.

If no MRL is available, an EPA RfD may be used to calculate a comparison value, because MRLs and RfDs are derived in a similar manner (4). The use of values other than MRLs for calculating screening values is left to the discretion and professional judgment of the health assessor.

Children usually constitute the most sensitive segment of the population for the ingestion of water, because their water ingestion rate per unit of body weight is greater than that of adults. An EMEG for a reference child is calculated by using a conservative water ingestion rate of 1 liter per day for a 10-kg child (5). For adults, a water EMEG is calculated by assuming a conservative water ingestion rate of 2 liters per day and a body weight of 70 kg. Water consumption surveys have indicated that a water consumption rate of 2 liters per day for an adult represents approximately the 90th percentile consumption rate (5). An illustrative example of EMEG calculations for a reference child and a reference adult are presented in Section A.5.

In addition to the ingestion of the contaminant in water, the assessor should also recognize the potential for inhalation of volatile organic compounds (VOCs) that escape from water used in the home. Experimental studies have demonstrated that VOCs are efficiently transferred from water to air, especially in showers where the water is heated and there is a large water-air interface. In model shower experiments, about 40-60% of trichloroethylene (a typical VOC) in water was volatilized to the air (6).

VOCs released into the air can then equilibrate with the air in the bathroom and eventually with the air in the rest of the house. Modeling has been used to calculate the concentration of VOCs in air in various parts of the house as a result of VOC release during indoor water use. These data, in combination with time-activity profiles of residents, have been used to estimate indoor air exposures to VOCs. For adults, it has been estimated that more than one half of the daily indoor air exposure to VOCs occurs in the shower stall, with an additional one third occurring in the bathroom. These models further predict that indoor air inhalation exposures could exceed ingestion exposures to VOCs in drinking water by 1.5- to 6-fold (7).

After being inhaled, VOCs can be absorbed by the respiratory epithelium and transported throughout the body by systemic blood circulation. Respiratory absorption of VOCs is influenced by the concentration in air, the duration of exposure, the blood/air partition coefficient, the solubility in various tissues, and physical activity which affects the ventilation rate and cardiac output (8). For chloroform, a typical VOC, respiratory absorption ranges from 49-77% (9). By contrast, oral absorption of chloroform has been reported to be approximately 100%(8).

While bathing and showering, skin contact with water leads to dermal absorption of VOCs and possibly of other contaminants in water. However, the available data are not adequate to support a quantitative estimation of dermal absorption for most contaminants (see Appendix D.5. - Dermal Exposure).

Recent experimental studies have demonstrated that a 10-minute shower yields an absorbed dose of chloroform that is equivalent to drinking 1.3 liters of water per day (10). The chloroform dose derived from showering was equally distributed between the inhalation and dermal exposure pathways (11).

Those studies indicate that inhalation (and dermal) exposures to VOCs in water can make a significant contribution to the total exposure dose. The magnitude of exposure varies depending on the frequency of showering and bathing, time spent indoors, air exchange rates in the bathroom and house, and other factors. Although a precise estimate of exposures by non-ingestion pathways will seldom be available, it may be estimated that non-ingestion exposures could yield a contaminant dose comparable to the ingestion dose. Therefore, when a VOC contaminant is present in a potable water supply, the water EMEG can be reduced by 50% to account for non-ingestion exposures.

A.2. SOIL EVALUATION GUIDES

Soil EMEGs are calculated using the following equation. As noted here, these EMEGs apply only to oral ingestion of soil.

$$EMEGs = \frac{MRL \times BW}{IR}$$

where,

EMEGs   = soil evaluation guide (mg/kg)

MRL   = minimal risk level (mg/kg/day)

BW   = body weight (kg)

IR   = soil ingestion rate (kg/day)

As discussed previously for water EMEGs, the chronic oral ingestion MRL for a chemical can be obtained from the ATSDR Toxicological Profile or from ATSDR's Hazardous Substances Data Management System (HAZDAT).

Many chemicals bind tightly to organic matter or silicates in the soil. Therefore, the bioavailability of a chemical depends on the media in which it is administered. Ideally, an MRL for deriving a soil EMEG should be based on an experiment in which the chemical is administered in soil. However, data from that type of study are seldom available. Therefore, it may be necessary to derive soil EMEGs from MRLs based on studies in which the chemical was administered in water or food.

Children are usually the most highly exposed segment of the population for ingestion of soil. Recent experimental studies have reported soil ingestion rates for children of about 200 mg per day (12, 13). Therefore, a conservative EMEG for a reference child is calculated using a soil ingestion rate of 200 mg per day for a 10-kg child.

Health assessors should also consider whether children who exhibit pica behavior have access to contaminated soil at a site. Soil ingestion in pica children greatly exceeds the soil ingestion rate for the normal population. As a useful reference point, an EMEG for a pica child can be calculated using a soil ingestion rate of 5,000 mg per day for a 10-kg child.

For sites where the only receptors for soil ingestion are adults, an EMEG is calculated using an adult body weight of 70 kg and an assumed soil ingestion rate of 100 mg per day. There are very few data on soil ingestion by adults, but preliminary studies suggest a soil ingestion rate in adults of 50-100 mg per day (14). Sample calculations for deriving soil EMEGs are shown in Section A.5.

The health assessor should also consider dermal absorption of soil contaminants and inhalation of dusts from contaminated soils. In both children and adults, the dose of a soil contaminant that results from oral ingestion is likely to exceed the dose resulting from dust inhalation (15). However, for contaminated dusts, special concern may be raised by chemicals that have specific toxic effects on the respiratory tract (e.g., chromium and lung cancer).

Little information exists on dermal absorption of chemicals from contaminated soil. Therefore, it is not feasible to quantitatively adjust soil EMEGs to account for dermal absorption of contaminants. However, the potential for exposure by this pathway should be acknowledged by the author of the health assessment.

It should also be noted that direct dermal contact with soil contaminants may provoke dermal sensitization reactions based on chemical reactivity or allergic sensitivity. Those types of sensitivity reactions result from direct skin contact with the chemical sensitizer and do not depend on dermal absorption of the contaminant. There is a high degree of variability in allergic sensitization reactions. The same dose of a skin sensitizer that causes severe dermatitis in one individual may elicit no response in another person. Therefore, sensitization reactions are not considered in the EMEGs for soil or other environmental media.

A.3. AIR EVALUATION GUIDES

EMEGs for inhalation exposures to airborne contaminants can be derived from the chronic inhalation MRLs presented in the ATSDR Toxicological Profiles or from the HAZDAT database. The inhalation MRLs are expressed in concentration units of milligrams/cubic meter or parts per million (ppm). Therefore, the air EMEG for a chemical is the same as its MRL, and no mathematical calculation is required as with water and soil EMEGs. For chemical substances that exist in a vapor form at standard temperature and pressure (STP), the value is given in ppm (volume basis); for substances that are solids at STP, the value is given in mg/cubic meter.

ATSDR's MRLs are derived for continuous, 24-hour-a-day exposures. In many instances, inhalation exposures from a site may be less than 24 hours a day. Therefore, the use of air EMEGs based on MRLs to assess those situations is a conservative approach to identifying air contaminants of potential health concern.

Inhalation MRLs are expressed as air concentrations rather than as a dose per unit of body weight. Therefore, it is not possible to calculate a range of air EMEG values as a function of exposure to contaminated air, and the same air EMEG value is used for all segments of the population.

For some chemicals, there may be experimental toxicity data in which the chemical was administered orally, but no data in which the chemical was administered by inhalation. Significant differences may exist in the toxicity of the chemical by ingestion, compared to inhalation exposure, because of differences in the absorption, metabolism, distribution, and site-specific toxicity of the chemical. Therefore, it is recommended that an air evaluation guide be derived only from an MRL based on an inhalation study. That approach is compatible with the Toxicological Profile protocols, which use only inhalation data to derive inhalation MRLs.

Dermal exposure to some air contaminants can also result in absorption through the skin. However, data are not likely to be available to quantitatively estimate exposures from this pathway. Therefore, it is not possible to adjust air EMEGs to account for this potential exposure pathway. Nevertheless, this potential exposure pathway should be acknowledged by the health assessor.

A.4. EDIBLE BIOTA

Aquatic Organisms

Human consumption of aquatic organisms may be an important source of exposure to an environmental contaminant. In the Ambient Water Quality Criteria documents (AWQC), EPA proposed values for water contaminants in order to protect humans who eat fish and other aquatic species harvested from a contaminated body of water (16). In calculating those water criteria, EPA used a bioconcentration factor (BCF) to account for contaminant uptake from ambient water by aquatic organisms. BCFs for a chemical that have been determined in experimental or field studies vary widely among species. In particular, BCFs for fish and shellfish vary greatly for many contaminants. Even within a given species, the BCF may vary according to age, size, lipid content, developmental stage, etc. Aquatic organisms may also bioaccumulate chemicals from contaminated sediment and food that they consume. Because of the numerous factors that influence BCFs, it is not possible to determine the water concentration of a contaminant that would protect humans who consume a variety of aquatic species living under varying conditions.

The EPA's AWQC assumes that an individual eats 6.5 grams of fish and shellfish per day. That figure represents the national average, and it may significantly over- or underestimate individual fish consumption.

Given the substantial uncertainties in BCFs and fish consumption rates, it is not feasible to propose water EMEGs based on human consumption of aquatic organisms. When preparing health assessments for sites where the consumption of aquatic organisms is a concern, the assessor should base the health assessment for biota consumption on actual measurements of the contaminant concentration in edible portions of the relevant aquatic species. The assessor should also consider the specific dietary habits of the potentially affected population. If that information is not available, the assessor should state that an acceptable evaluation of this exposure pathway cannot be made without the information.

Terrestrial Organisms

Humans may also be exposed to environmental contamination through consumption of terrestrial plants or animals that have been grown or raised in contaminated areas. Plants or animals may bioaccumulate chemical contaminants from soil, water, or air. The database of information for the uptake of contaminants from the environment by terrestrial biota is very limited. Therefore, it is not currently possible to propose methodology that can be used to derive EMEGs for soil, water, or air based on the consumption of contaminated terrestrial biota.

Action Levels and Tolerance Levels

Action levels for poisonous or deleterious substances are established by the Food and Drug Administration (FDA) to control levels of contaminants in human foods and animal feed. Tolerance levels, established by EPA, are maximum allowable levels of pesticide residues in or on raw agricultural products and in processed food. A discussion of action levels is included in Appendix B.11. Tolerance levels are found in the Code of Federal Regulations, title 40, part 193; title 40, part 180; and title 21, part 109.30. Because of the complexity of those regulations (each chemical can have tolerance levels for dozens of distinct food products), tolerance levels have not been included in this document. FDA Action Levels and EPA Tolerance Levels may underestimate food consumption for certain groups—for example, sport fishermen and backyard gardeners. Therefore, those values should not be used to select contaminants of concern unless the assumed food consumption rates are determined to be appropriate for the population being evaluated.

A.5. CALCULATING EMEG VALUES

Calculations for deriving EMEGs for 1,1-dichloroethene are shown here under Illustrative Calculations. These values are calculated using the chronic MRLs presented in the HAZDAT database. Because MRLs are subject to periodic updates, health assessors should ensure that they are using the most current MRL value when calculating an EMEG. As a reference source, ATSDR will maintain a current list of MRLs and EMEG values in HAZDAT. A sample printout of the EMEG field from HAZDAT is shown in Figure 1. The EMEG values in this table have been rounded off to one significant figure.

As previously discussed, EMEGs are based on the noncarcinogenic toxic effects of the chemical. In the HAZDAT database, chemicals are marked to indicate that they have been identified as having carcinogenic toxicity by either the National Toxicology Program, the International Agency for Research on Cancer, or the EPA. For chemicals with carcinogenic toxicity, it may be necessary to develop an alternative comparison value that considers the chemical's carcinogenic potential (see Section 5.6.).

EMEGs are explicitly presented as a range of values, rather than as a single, discrete number. This range is based on variable exposures to a contaminated medium that could occur among different individuals in the population. Health assessors should select an EMEG that will be protective of the most highly exposed segment of the population at a site. As an aid to health assessors, several important reference points, which apply to different segments of the population, are indicated. However, if sufficient information is available, health assessors may derive an alternate EMEG value based on site-specific exposure information.

Figure 1.

ATSDR HAZDAT 1.0 EMEG RANGE**			
CAS#: 000075-35-4	1,1-DICHLOROETHENE *		
RECEPTOR POPULATION	WATER EMEG (MG/L)	SOIL EMEG (MG/KG)	AIR EMEG (PPM)
REFERENCE ADULT	0.3	6000	0.02
REFERENCE CHILD	0.09	400	0.02
PICA CHILD		20	

* - EVIDENCE OF CARCINOGENIC TOXICITY. SEE THE ATSDR TOXICOLOGICAL PROFILE FOR FURTHER INFORMATION.

** EMEG RANGE IS FROM LOWEST TO HIGHEST CONCENTRATION SHOWN FOR EACH MEDIUM.

EMEGS ARE TO BE USED ONLY FOR THE SELECTION OF CONTAMINANTS OF CONCERN AS DESCRIBED IN THE ATSDR HEALTH ASSESSMENT GUIDANCE MANUAL. THEY SHOULD NOT BE USED AS PREDICTORS OF ADVERSE HEALTH OUTCOMES OR FOR ANY REGULATORY PURPOSE.

ILLUSTRATIVE CALCULATIONS

Water EMEG

$$EMEGw = \frac{MRL \times BW}{IR}$$

EMEGw   = Water EMEG (mg/L)

MRL   = Minimal Risk Level (mg/kg/day)

BW   = Body weight (kg)

IR   = Ingestion Rate (L/day)

Reference Child:

$$EMEGw = \frac{(0.009 \text{ mg/kg/day}) \times (10 \text{ kg})}{1L/day}$$

EMEGw = 0.09 mg/L

Reference Adult:

$$EMEGw = \frac{(0.009 \text{ mg/kg/day}) \times (70 \text{ kg})}{(2L/day)}$$

EMEGw = 0.315 mg/L

Soil EMEG

$$EMEGs = \frac{MRL \times BW}{IR}$$

EMEGs   = Soil EMEG (mg/kg)

MRL   = Minimal Risk Level (mg/kg/day)

BW   = Body weight (kg)

IR   = Ingestion Rate (kg/day)

Pica Child:

$$EMEGs = \frac{(0.009 \text{ /mg/kg/day}) \times (10kg)}{5,000 \times 10^{-6} \text{ kg/day}}$$

EMEGs = 18 mg/kg

Reference Child:

$$EMEGs = \frac{(0.009 \text{ mg/kg/day}) \times (10 \text{ kg})}{200 \times 10^{-6} \text{ kg/day}}$$

EMEGs = 450 mg/kg

Reference Adult:

$$EMEGs = \frac{(0.009 \text{ mg/kg/day}) \times (70 \text{ kg})}{100 \times 10^{-6} \text{ kg/day}}$$

EMEGs = 6300 mg/kg

Air EMEG

$$EMEG_A = \text{Inhalation MRL}$$

EMEG$_A$ = Air EMEG (ppm)

MRL   = Minimal Risk Level (ppm)

EMEG$_A$ = 0.02 ppm

A.6. References

1. Federal Register, Vol. 51, No. 185, pages 34014-34025, Wednesday, September 24, 1986.

2. National Research Council, Complex mixtures. Washington DC: National Academy Press, 1988.

3. U.S. Department of Health and Human Services. Fifth Annual Report on Carcinogens-1989. Washington, DC: U.S. Government Printing Office, 1989.

4. Barnes DG and Dourson M. Reference dose (RfD): Description and use in health risk assessments. Regulatory Toxicology and Pharmacology 1988; 8:471-86.

5. EPA Office of Health and Environmental Assessment. Exposure factors handbook.

Washington, DC: Environmental Protection Agency, March 1990; EPA/600/8-89/043.

6. Andelman JB. Inhalation exposure in the home to volatile organic contaminants of drinking water. The Science of the Total Environment 1985; 47:443-60.

7. McKone TE. Human exposure to volatile organic compounds in household tap water: the inhalation pathway. Environ Sci Technol 1987; 21:1194-201.

8. ATSDR Division of Toxicology. Toxicological Profile for Chloroform. Atlanta, GA: Agency for Toxic Substances and Disease Registry, January 1989; TP-88-033.

9. EPA Office of Water Regulations and Standards. Ambient water quality criteria for chloroform. Washington, DC: EPA/440/5-80-033.

10. Jo WK, Weisel CP, Lioy PJ. Chloroform exposure and the health risk associated with multiple uses of chlorinated tap water. Risk Analysis 1990; 10(4):581-85.

11. Jo WK, Weisel CP, Lioy PJ. Routes of chloroform exposure and body burden from showering with chlorinated tap water. Risk Analysis 1990; 10(4):575-80.

12. Calabrese EJ, et al. How much soil do young children ingest: an epidemiologic study. Regulatory Toxicology and Pharmacology 1989; 10:123-37.

13. Davis S, et al. Quantitative estimates of soil ingestion in normal children between the ages of 2 and 7 years. Archives of Environmental Health 1990; 45:112-22.

14. Calabrese EJ, et al. Preliminary adult soil ingestion estimates: results of a pilot study. Regulatory Toxicology and Pharmacology 1990; 12:88-95.

15. Hawley JK. Assessment of health risk from exposure to contaminated soil. Risk Analysis 1985; 5:289-302.

16. Federal Register, Vol. 45, No. 231, pages 79318-79. Friday, November 28, 1980.

Health guidelines and standards for maximum permissible concentrations of potentially hazardous materials in water, air, and food have been developed by both federal and state agencies. **Standards** are developed by regulatory agencies and are legally enforceable (e.g., Maximum Contaminant Levels [MCLs]). **Health guidelines** are developed by various agencies as concentrations or doses that are likely to be protective of human health (e.g., health advisories and reference doses) and are not legally enforceable. This appendix is a compilation of health guidelines and standards useful in performing health assessments.

For each guideline or standard discussed, the following information is provided: (1) the definition of the standard/guideline; (2) the applicability or intended use of the standard/guideline; and (3) the legal citation or reference source. Because these standards and guidelines are frequently revised, any published table of values would soon become outdated. Therefore, numerical values for these standards and guidelines are not presented in this manual. ATSDR will prepare a supplement to this guidance manual that will contain an updated list of values. The supplement will be periodically revised in order to ensure that the values are current. In addition, current values for many of the standards and guidelines can be obtained from EPA's Integrated Risk Information System (IRIS), or from the reference sources listed at the end of each section.

B.1. SUBCHRONIC AND CHRONIC REFERENCE DOSES (RFDs)

Definition. A reference dose (RfD) is an estimate (uncertainty spanning perhaps an order of magnitude) of a daily exposure (mg/kg/day) to the general human population (including sensitive subgroups) that is likely to be without an appreciable risk of deleterious effects during a lifetime of exposure. The RfD is a benchmark dose derived from the No Observed Adverse Effect Level (NOAEL) or Lowest Observed Adverse Effect Level (LOAEL) by application of uncertainty factors that reflect various types of data used to estimate RfDs and an additional modifying factor, which is based on a professional judgment of the entire database of the chemical. An oral RfD is determined by the following equation:

$$RFD = \frac{NOAEL}{UF \times MF}$$

where,

UF = uncertainty factor
MF = modifying factor

The RfD is expressed in units of milligrams of contaminant per kilograms body weight per day (mg/kg/day).

An uncertainty factor (UF) of 10 is used when reference doses are based on experimental studies using prolonged exposure to average healthy humans. This factor is intended to account for variability in sensitivity among members of the human population. An additional tenfold UF is used when RfDs are based on experimental studies using long-term exposure to animals. This factor is intended to account for the uncertainty involved in extrapolating animal data to humans. Another tenfold factor is used when RfDs are based on a LOAEL instead of a NOAEL to account for uncertainty in extrapolating NOAELs from LOAELs. An additional tenfold safety factor is used when extrapolating from less than chronic results on experimental animals. This factor is intended to account for the uncertainty in extrapolating from less than chronic NOAELs to chronic NOAELs.

A modifying factor (MF) is another uncertainty factor, ranging from greater than 0 to 10, to reflect professional assessment of scientific uncertainties not explicitly covered by the UF (e.g., the completeness of the overall database and the number of species tested). The default value for MF is 1.

The subchronic RfD is an estimate (in mg/kg/day) of an exposure level that would not be expected to cause adverse effects when exposure occurs during a limited time interval. Subchronic values are determined from animal studies with durations of 30-90 days. Subchronic human exposure information is usually derived from occupational exposures and accidental acute exposures.

Applicability/Intended Use. The RfD is based on the assumption that thresholds exist for certain toxic effects such as cellular necrosis, but may not exist for other toxic effects such as carcinogenicity. RfDs can also be derived for the noncarcinogenic health effects of compounds that are also carcinogens. Therefore, it is essential to refer to other sources of information concerning the carcinogenicity of this substance. Footnotes identify animal species used in calculating the RfD, RfDs based on route-to-route extrapolation, and RfDs that are in units different from those specified in the column headings.

Reference/Legal Citation. RfDs are listed in EPA's IRIS database and in Health Effects Assessment Documents, and summarized in Health Effects Assessment Summary Tables (HEAST) OSWER (OS-230), ORD (RD-689), OERR 9200.6-303-(89-4), October 1989.

B.2. CANCER SLOPE FACTOR

Definition. In evaluating the potential human carcinogenicity of chemicals, EPA uses the approach given in "Guidelines for Carcinogenic Risk Assessment" (51 FR 33992, September 24, 1986). Determining the carcinogenic potential of a chemical is a two-step process. The first is a qualitative evaluation made by considering all the available information relevant to carcinogenicity and judging the quality of the information. This is termed the weight-of-evidence assessment. This assessment is used to categorize the chemicals. The second step involves performing a quantitative assessment to define the relationship between dose and the likelihood of an increase in carcinogenic effect over that seen in controls (the dose-response assessment).

When the weight-of-evidence classification is determined, studies in humans and animals are evaluated separately and labeled according to the following criteria (51 FR 33992). The evidence for carcinogenicity from studies in humans is classified as:

- **sufficient evidence** -- there is a causal relationship between the agent and human cancer;

- **limited evidence** -- a causal interpretation is credible, but alternative explanations, such as chance, bias, or confounding, could not adequately be excluded;

- **inadequate evidence** -- one of two conditions prevail:
  (a) there are few pertinent data; or
  (b) the available studies, while showing evidence of association, do not exclude chance, bias, or confounding and therefore a causal interpretation is not credible;

- **no data** -- no data are available; and

- **no evidence** -- no association was found between exposure and an increased risk of cancer in well-designed and well-conducted independent analytical epidemiologic studies.

The evidence for carcinogenicity from studies in animals is classified as:

- **sufficient evidence** -- there is an increased incidence of malignant tumors or combined malignant and benign tumors: (a) in multiple species or strains; (b) in multiple experiments (e.g., with different routes of exposure or different dose levels; or (c) to an unusual degree in a single experiment (i.e., high incidence, unusual site or type of tumor, or early age at onset);

- **limited evidence** -- the data suggest a carcinogenic effect, but are limited because: (a) the studies involve a single species, strain, or experiment and do not meet criteria for sufficient evidence; (b) the experiments are restricted by inadequate dosage levels, inadequate duration of exposure, inadequate period of follow-up, poor survival, inadequate numbers of animals, or inadequate reporting; or (c) an increase is apparent in the incidence of benign tumors only;

- **inadequate evidence** -- because of major qualitative or quantitative limitations, the studies cannot be interpreted as showing either the presence or absence of a carcinogenic effect;

- **no data** -- no data are available;

- **no evidence** -- there is no increased incidence of neoplasms in at least two well-designed and well-conducted animal studies in different species.

After the weight-of-evidence for human and animal studies is evaluated, the matrix below is used to categorize a potential carcinogen in the following groups:

A. Human carcinogen
B. Probable human carcinogen
C. Possible human carcinogen
D. Not classifiable as to human carcinogenicity
E. Evidence of noncarcinogenicity in humans

Group B is subdivided into two groups. Group B1 is used for chemicals for which there is limited evidence of carcinogenicity from epidemiologic studies. Group B2 is used to categorize chemicals for which there is sufficient evidence of carcinogenicity in animals, but inadequate evidence or no data from epidemiologic studies.

EPA expresses toxicity values for carcinogenic effects as slope factors. The slope factor is usually, but not always, the upper 95th percentile confidence limit of the slope for the

WEIGHT-OF-EVIDENCE CLASSIFICATION FOR CARCINOGENS					
	Animal Evidence				
Human Evidence	sufficient	limited	inadequate	no data	no evidence
sufficient	A	A	A	A	A
limited	B1	B1	B1	B1	B1
inadequate	B2	C	D	D	D
no data	B2	C	D	D	E
no evidence	B2	C	D	D	E

dose-response curve and is expressed as $(mg/kg/day)^{-1}$. If the extrapolation model selected is the linearized multistage model, this value is also known as the q1*. When data permit, slope factors listed in IRIS are based on absorbed doses, although many of them have been based on administered doses.

Assuming a continuous, lifetime exposure to a carcinogen, the risk (R) and the dose (D) in mg/kg/day are related by the equation:

$$R = q1^* \times D$$

This equation is valid only at low risk levels (i.e., below estimated risks of 0.01).

EPA has developed separate q1* values for ingestion and inhalation pathways. No values currently exist for dermal absorption.

Applicability/Intended Use. Slope factors are from EPA's IRIS database and Health Effects Assessment Documents, summarized in HEAST. Slope factors are usually derived from animal experiments that involve exposure to a chemical by a single route of exposure (e.g., ingestion or inhalation). There may be qualitative or quantitative differences in the carcinogenicity of a chemical depending on the route of exposure. Therefore, a cancer slope factor derived from one route of exposure should not be applied to a different route of exposure unless there is adequate justification for the extrapolation.

Reference/Legal Citation. Slope factors and carcinogen classes are listed in EPA's IRIS database and in Health Effects Assessment Documents, and summarized in Health Effects Assessment Summary Tables (HEAST) OSWER (OS-230), ORD (RD-689), OERR 9200.6-303-(89-4), October 1989. EPA drinking water health advisories provide an additional source for carcinogen classes for some chemicals. Carcinogenic assessment is discussed in Guidelines for Carcinogenic Risk Assessment. 51 *Federal Register* 33992, September 24, 1986.

B.3. EPA DRINKING WATER HEALTH ADVISORIES

Definition. Health advisories (HAs) provide the level of a contaminant in drinking water at which adverse noncarcinogenic health effects would not be anticipated with a margin of safety. Drinking water concentrations are developed to predict acceptable 1-day, 10-day, and longer-term (approximately 7 years) exposure levels for both adults and children when data on a NOAEL or LOAEL exist from animal or human studies. Short-term HAs are intended to be used for short-term exposures such as spills and accidents. Lifetime HAs represent that portion of an individual's total exposure to a chemical that is attributed to drinking water, and that is considered protective of noncarcinogenic health effects during a lifetime (70 years) exposure. EPA has developed Maximum Contaminant Level Goals (MCLGs) and MCLs from lifetime HAs.

The general formula for estimating a drinking water health advisory concentration is:

$$HA = \frac{NOAEL \text{ or } LOAEL \times BW}{UF \times IR}$$

where,

HA = health advisory (μg/L)

NOAEL = no observed adverse effect level (μg/kg/day)

LOAEL = lowest observed adverse effect level (μg/kg/day)

BW = assumed body weight of a child (10 kg) or an adult (70 kg)

UF = uncertainty factor

IR = assumed daily water ingestion rate of a child (1 L/day) or for an adult (2 L/day)

Short-term HAs are calculated by analyzing a study of appropriate duration (1-day, 10-day, or subchronic) and selecting the NOAEL or LOAEL. The drinking water equivalent level (DWEL) is a lifetime exposure level specific for drinking water (assuming that all exposure is from that medium) at which adverse, noncarcinogenic health effects would not be expected to occur. The DWEL health advisory is unchanged for class A and B carcinogens. Lifetime health advisories (LHA) are derived from DWELs for noncarcinogens. For noncarcinogenic organic compounds, LHAs are 20% of the DWEL; for noncarcinogenic inorganic compounds, LHAs are 10% of the DWEL. For Class C carcinogens, the LHA is divided by an additional factor of 10. The LHA is not determined for class A and B carcinogens.

When sufficient information is available, the water concentration corresponding to a carcinogenic risk of 10^{-4} may be calculated.

Health advisories also contain a somewhat detailed description on chemical properties, pharmacokinetics, and health effects in humans and animals.

Applicability/Intended Use. HAs are not legally enforceable standards; they are not issued as an official regulation, and they may or may not lead ultimately to the issuance of a national standard or MCL. Because MCLs take into account occurrence, relative source contribution factors, treatment technologies, monitoring capability, and costs in addition to health, it is more than likely that any resulting MCL would differ from the strictly health-based HA.

The existence of an HA does not condone the presence of contaminants in drinking water, but rather provides useful information to assist in setting control priorities in cases when they have been found. The Office of Drinking Water HAs usually do not consider the health risk resulting from possible synergistic effect of other chemicals in drinking water, food, and air.

Reference/Legal Citation. 50 FR 46936, November 13, 1985. The EPA Office of Drinking Water Health Advisory Documents. Health advisory values may be reevaluated and calculated without publishing new health advisory documents. EPA provides a monthly summary, *Drinking Water Regulations and Health Advisories* by the Office of Drinking Water [(202) 382-7571], that contains revised and draft values.

B.4. SAFE DRINKING WATER ACT MAXIMUM CONTAMINANT LEVELS AND GOALS

Definition. The Safe Drinking Water Act (SDWA) establishes national primary drinking water regulations in the form of MCLs. MCLs are enforceable drinking water regulations that are protective of public health to the "extent feasible." National primary drinking water regulations apply to all public water systems including community water systems and transient and nontransient noncommunity water systems.

An MCL is the maximum permissible level of a contaminant in water that is delivered to the free-flowing outlet of the ultimate user of a public water system. The only exception to this rule involves measurement of turbidity; in that case, the maximum permissible level is measured at the point of entry into the distribution system. Contaminants added to the water by the user, except those resulting from corrosion of piping and plumbing caused by water quality, are excluded from this definition.

By law, MCLs are monitored on a prescribed schedule (frequency) and by means of a specified analytical method. Legal violation of an MCL is not determined or based on the results of a single sample; rather, it is based on a series of samples taken over the prescribed monitoring period. MCLs are set by EPA as close to MCLGs as is "feasible" with the use of the best technology, treatment techniques, and other means which the EPA Administrator finds, after examination for efficacy under field conditions and not solely under laboratory conditions, are available (taking cost into consideration).

MCLGs (formerly known as Recommended Maximum Contaminant Levels - RMCLs) are drinking water health goals. MCLGs are to be set at a level at which, in the EPA Administrator's judgment, "no known or anticipated adverse effect on human health occurs and which allows an adequate margin of safety." The Administrator must consider the possible impact of synergistic effects, long-term and multi-stage exposures, and the existence of more susceptible groups in the population. When there is no safe threshold for a contaminant (including Group A and B carcinogens), the MCLG should be set at zero.

PMCLGs are Proposed Maximum Contaminant Level Goals.

Applicability/Intended Use. MCLs are the heart of the national primary drinking water regulations, and have been issued by the EPA under the authority of the Safe Drinking Water Act (SDWA). Drinking water standards in the United States were originally promulgated in 1914; they were reissued or revised in 1925, 1942, 1946, and 1962. While the 1914 drinking water standards were concerned solely with bacteriologic quality, the 1925 standards and those of following years include maximum permissible limits for chemical constituents. Although the 1962 U.S. Public Health Service Drinking Water Standards were replaced in 1975 (effective in 1977) by national interim primary drinking water regulations, many of the original maximum permissible limits from 1962 were adopted as MCLs. MCLs are now periodically proposed or reevaluated, and are set as close to MCLGs as is feasible.

MCLs are deemed protective of public health (considering the availability and economics of water treatment technology) over a lifetime (70 years) at an exposure rate of 2 liters water per day. MCLs are dynamic values and subject to change as water treatment technologies and their economics evolve and/or as new toxicologic information warrants.

Besides their primary use as quality standards for public water supplies, MCLs are useful in evaluating water quality data from private water supplies (generally water wells) for determining potability. When applying MCLs to private water supplies, however, one must remember that their development involved aspects beyond those bearing strictly on health. It must also be remembered that MCLs are not intended to apply to single sample results, or to results from source water samples. To be reasonably applied, data must have originated from the MCL-specified analytical procedure or procedures.

MCLGs and PMCLGs do not have legal application to public water supplies, nor do they carry any legal authority under SDWA. When MCLGs are developed, health advisories will be developed or reevaluated.

MCLGs and PMCLGs as well as MCLs are useful as screening parameters for determining potability of non-public water supplies (private water well supplies). For that type of application, MCLGs and PMCLGs may be more applicable than MCLs because they are strictly health based. Application in this manner is reinforced by provisions of the Superfund Amendments and Reauthorization Act of 1986. The Act requires remedial actions to attain at least a degree of cleanup and control of further release in order to protect human health and the environment. The Act now requires at least attaining MCLGs when appropriate.

Reference/Legal Citation. National Primary Drinking Water Regulations (40 CFR 141) - Safe Drinking Water Act (42 U.S.C. 300g-1, 300g-3, 300j-4, and 300j-9; enacted by PL 93-523, December 16, 1974; last amended June 19, 1986, by PL 99-339); and the National Primary Drinking Water Regulations - Background Document.

B.5. SAFE DRINKING WATER ACT SECONDARY MAXIMUM CONTAMINANT LEVELS

Definition. National Secondary Drinking Water Regulations were established under Section 1412 of the Safe Drinking Water Act as amended (42 U.S.C. 300g-1). These regulations control drinking water contaminants that affect the aesthetic qualities of water, and they are related to public acceptance of water. Health implications, as well as aesthetic problems, may occur at higher concentrations of contaminants.

Applicability/Intended Use. The regulations are not federally enforceable and are intended only as guidelines for the states. Higher or lower levels may be developed by the states based on local conditions. The Secondary Drinking Water Regulations recommend that these contaminants be monitored.

Reference/Legal Citation. The regulations, referred to as Secondary Maximum Contaminant Levels, are published in the Code of Federal Regulations (40 CFR 143).

B.6. WATER QUALITY CRITERIA

Definition. Water quality criteria (WQC) are the recommended maximum permissible pollutant concentrations protective of aquatic organisms and human health (i.e., when people participate in recreational activities or otherwise contact water). For noncarcinogens, WQC are based on acceptable daily intakes (ADIs are now called reference doses [RfD]). In establishing water quality criteria for priority pollutants, EPA used the following relationship:

$$Cw = \frac{BW \times ADI}{IR + If \times BCF}$$

where,

Cw = water quality criteria level (mg/L);
BW = body weight (70 kg);
ADI = acceptable daily intake (mg/kg/day);
IR = water ingestion rate (2 L/day);
If = daily fish consumption rate (0.0065 kg/day);
BCF = water-to-fish bioconcentration factor (L/kg).

Before being used in this formula, ADI values should be adjusted to reflect background daily intake from non-water-related sources. When ADI data were not available, Threshold Limit Values (TLVs) or animal inhalation studies were used to generate a criterion. In some cases, organoleptic (taste and odor) properties form the basis for the criterion. An organoleptic criterion makes no statement concerning adverse health effects. For specific information, the background documents should be consulted.

For carcinogenic compounds, WQC are based on lifetime cancer risk. Calculation of WQC is based on the following formula:

$$Cw = \frac{BW \times LRF}{CSF \times (IR + If \times BCF)}$$

where,

Cw = water quality criteria level (mg/L)
BW = body weight (70 kg);
LRF = lifetime risk factor (10^{-5}, 10^{-6}, or 10^{-7});
CPF = cancer slope factor $(mg/kg/day)^{-1}$;
IR = water ingestion rate (2 L/day)
If = daily fish consumption rate (0.0065 kg/day);
BCF = water-to-fish bioconcentration factor (L/kg).

Applicability/Intended Use. WQC values are currently being revised. Although they are not federally enforceable, some states have adopted them as standards for specified water uses. WQC present scientific data and guidance on the environmental and human health effects of pollutants, which can be useful for deriving regulatory requirements based on considerations of water quality impact. Standards that have been based on WQC include the water quality-based effluent limitations under Section 302, water quality standards under Section 303, and toxic pollutant effluent standards under Section 307. WQC do not reflect considerations of economic or technologic feasibility. The criteria for the Section 307(a)(1) toxic pollutants are based solely on the effect of a single pollutant. WQC that are designated for the protection of human health or for water supply have application as screening parameters for water quality determinations, including potability. However, the criteria were not strictly intended for this use and must be used with caution. The criteria were developed in part around the ultimate protection of surface water bodies as source water for public water supplies, but were not intended to be applied directly to drinking water itself.

Because the latest human health criteria consider some contribution from contaminated aquatic organisms, the direct application of these criteria to water supply sources, including groundwater sources used for drinking water, was not intended. However, when groundwater is used for aquaculture, many of the human health criteria would apply. In order that human health water quality criteria (1980) could be used for assessing water supplies and groundwater, they have been adjusted to "drinking water only" by EPA's Office of Emergency and Remedial Response and Office of Solid Waste and Emergency Response. The contaminant contribution from ingesting contaminated aquatic organisms was subtracted. The adjusted water quality criteria are not official EPA Water Quality Criteria.

Reference/Legal Citation. 1) Clean Water Act. 2) Water Quality Criteria Documents for each of the 65 toxic pollutants, EPA, Office of Water Regulations and Standards, Criteria and Standards Division. 3) Water Quality Criteria Documents; Availability of Summaries for 64 Toxic Pollutants/Categories, *Federal Register*, 45, No. 231:79318-79379, November 28, 1980. 4) Adjusted Water Quality Criteria—For Drinking Water Only, *Guidance on Feasibility Studies Under CERCLA*, Office of Emergency and Remedial Response and Office of Solid Waste and Emergency Response, U.S. EPA, June 1985. 5) Water Quality Criteria. U.S. EPA, 1968 ("Green Book"). 6) Water Quality Criteria 1972. U.S. EPA ("Blue Book"). 7) Quality Criteria for Water. U.S. EPA, 1976 ("Red Book").

Note: The WQC Documents Summary supersedes all previous WQC for human health for the 64 contaminants summarized. When WQC are not superseded, the WQC in the Red Book and in the Blue Book are still in effect. WQC documents are an excellent source for chemical properties, pharmacokinetics, and acute and chronic

health effects in humans and animals as well as for carcinogenicity data.

B.7. OSHA PERMISSIBLE EXPOSURE LIMITS (PELs)

Definition. The Occupational Safety and Health Act (OSHA) of 1970 provides for safe and healthful working conditions for working men and women. This is accomplished by setting occupational safety and health standards and by providing research, information, and training in the field of occupational safety and health. The OSHA standard is known as the permissible exposure limit (PEL).

Applicability/Intended Use. OSHA has set PELs for certain airborne contaminants in the workplace based on health criteria and technical feasibility. They are designed to assure, to the extent feasible, that no employee suffers impairment of health or functional capacity even if he/she is regularly exposed to a toxic material throughout working life. PELs are usually listed as 8-hour time-weighted averages (TWA). The level may be exceeded, but the sum of the exposure levels averaged over 8 hours must not exceed the limit. In some cases, ceiling and peak levels are listed in place of, or in addition to, the 8-hour TWA. Ceiling values cannot be exceeded at any time. During a designated time period, they may reach, but never exceed, a peak level.

The short-term exposure limit (STEL) is a 15-minute time-weighted average which should not be exceeded at any time during a workday even if the 8-hour TWA is within the PEL. Exposures at the STEL should not exceed 15 minutes and should not be repeated more than four times per day. There should be at least a 60-minute interval between successive exposures at the STEL. A STEL is recommended only in cases in which toxic effects have been reported from high short-term exposures in either animals or humans. It is not a separate, independent exposure limit, but rather a supplement to the PEL.

It is important to understand that PELs apply to healthy adult employees working 40-hour weeks, and do not apply to the general population—including children, the elderly, and the sick—which may be subject to continuous environmental exposure.

Reference/Legal Citation. *Federal Register* 54, No. 12, pages 2332-2983. January 19, 1989.

The American Conference of Government Industrial Hygienists annually prepares a list of recommended air standards for occupational exposures. This list of standards (Threshold Limit Values) is not reproduced here because most of the standards are identical to the OSHA PELs. A list of the recommended standards and the rationale for their derivation can be obtained from the following references.

American Conference of Governmental Industrial Hygienists, *Threshold Limit Values and Biological Exposure Indices*, ACGIH, 6500 Glenway Ave., Bldg D-7, Cincinnati, Ohio 45211.

American Conference of Governmental Industrial Hygienists, *Documentation of the Threshold Limit Values and Biological Exposure Indices, Fifth Edition*, ACGIH, 6500 Glenway Ave., Bldg D-7, Cincinnati, Ohio 45211.

B.8. NIOSH RECOMMENDED EXPOSURE LIMITS (RELs)

Definition. Under the authority of the Occupational Safety and Health Act (OSHA) of 1970 (Public Law 91-596), the National Institute for Occupational Safety and Health (NIOSH) develops and periodically revises recommendations or limits of exposure to potentially hazardous substances or conditions in the workplace. These recommendations are then published and transmitted to OSHA and

the Mine Safety and Health Administration for use in promulgating legal standards.

Applicability/Intended Use. NIOSH has published RELs for airborne contaminants in the workplace. The RELs are developed for 8- or 10-hour time-weighted averages (TWA) or for ceiling levels. A ceiling level should not be exceeded during any part of the working exposure.

NIOSH has also published Immediately Dangerous to Life and Health levels (IDLH), which represent the maximum concentration from which one could escape within 30 minutes without escape-impairing symptoms or irreversible health effects.

NIOSH identifies chemicals that could be treated as occupational carcinogens using a classification outlined in 29 CFR 1990.103 which, in part, states: "Potential occupational carcinogen means any substance, or combination or mixture of substances, which causes an increased incidence of benign and/or malignant neoplasms, or a substantial decrease in the latency period between exposure and onset of neoplasms in humans or one or more experimental mammalian species as the result of oral, respiratory or dermal exposure, or any other exposure which results in the induction of tumors at a site other than the site of administration. This definition also includes any substance which is metabolized into one or more potential occupational carcinogens by mammals."

Reference/Legal Citation. 1) NIOSH Pocket Guide to Chemical Hazards, Public Health Service, Centers for Disease Control, National Institute for Occupational Safety and Health. June 1990. 2) NIOSH Recommendations for Occupational Safety and Health Standards 1988, *Morbidity and Mortality Weekly Report*, August 26, 1988.

B.9. NATIONAL AMBIENT AIR QUALITY STANDARDS (NAAQS)

Definition. National Ambient Air Quality Standards (NAAQS) are set under Section 109 of the Clean Air Act (CAA) for any pollutants which, if present in air, might endanger the public health (primary standards) or public welfare (secondary standards). In the development of primary standards, all sources of the pollutant that contribute to the health risk are considered. The standards must allow for an adequate margin of safety and must consider the nature and severity of the health effects of each contaminant, the most sensitive group of individuals at risk, and the degree of uncertainty of the scientific evidence. The CAA does not require EPA to consider economic or technical feasibility of implementing the standards.

Applicability/Intended Use. The NAAQSs are not directly enforceable; rather, they establish ceilings that are not to be exceeded in an area in which the source or sources of the pollutant are located. Thus, the standards determine restrictions on new sources, and the degree of control to be imposed on existing sources. In effect, these controls determine if a new facility can be built in a given region and the type of pollution abatement systems that new and existing facilities must install. Standards can be promulgated as annual maximums, annual geometric means, annual arithmetic means, or for other time periods that vary from 1 hour to 1 year, depending on the pollutant

Reference/Legal Citation. 40 CFR 50.4, 50.6, 50.8, and 50.9 - 50.12.

B.10. NATIONAL EMISSION STANDARDS FOR HAZARDOUS AIR POLLUTANTS (NESHAPs)

Definition. National Emission Standards for Hazardous Air Pollutants (NESHAPs) are set by EPA under Section 112 of the Clean Air Act for dangerous pollutants not covered by NAAQS because they are not emitted by a wide range of sources. Exposure to these pollutants might result in an increase in mortality or an increase in serious irreversible illness or incapacitating reversible illness. After listing pollutants, EPA must establish standards within 180 days for new and existing sources.

Applicability/Intended Use. EPA must consider an ample margin of safety but may also consider feasibility and costs in setting emission standards for these pollutants. New sources must comply with NESHAPS at start-up; existing sources must comply within two years.

Reference/Legal Citation. 40 CFR 61.

B.11. FDA ACTION LEVELS

Definition. Action levels are maximum allowable levels of poisonous and deleterious substances in human food and in animal feed. Established by the U.S. Food and Drug Administration, action levels exist for approximately 23 toxic substances and are expressed in parts per million (ppm).

Tolerance levels are maximum allowable levels of pesticide residues in or on raw agricultural products and in processed food. Established by EPA, tolerance levels are expressed in parts per million (ppm) (mg of contaminant per kg of foodstuff). Because tolerance levels are established for chemicals and foodstuffs with already established action levels, the tolerance level replaces the action level.

Applicability/Intended Use. Tolerance levels have not been included in this appendix because of the quantity of chemicals and foodstuffs associated with each chemical. Tolerance levels are located in several parts of the *Code of Federal Regulations (CFR)*. Tolerance levels for contaminants in processed foods are found in title 40, part 185, of the CFR. Tolerance levels for residues in raw agricultural products are located in title 40, part 180, of the CFR. Tolerance levels for polychlorinated biphenyls are listed separately, in title 21, part 109.30, of the CFR.

Reference/Legal Citation. The U.S. Food and Drug Administration, *Action Levels for Poisonous or Deleterious Substances in Human Food and Animal Feed.*

B.12. RCRA APPENDIX VIII AND IX, SUPERFUND TARGET SUBSTANCES, AND CLEAN WATER ACT (CWA) PRIORITY POLLUTANT COMPOUNDS

Definition. Under Section 307(a)(1) of the Clean Water Act, EPA established a priority list of toxic pollutants for which federally enforceable discharge limits were set. Effluent limitations for those priority pollutants are based on the best available technology that is economically achievable for the applicable category or class of point source creating the discharge.

The Administrator of EPA is authorized to add substances to or remove them from the list. Additions to or deletions from the list must take into account the toxicity of the pollutant, its persistence, degradability, the usual or potential presence of affected organisms in any waters, the importance of the affected organisms, and the nature and extent of the effect of the toxic pollutant on such organisms.

Applicability/Intended Use. This table identifies the statutes under which various chemicals are regulated and is provided as a source of information.

Reference/Legal Citation. 40 CFR 261, Appendix VIII and 40 CFR 264, Appendix IX. EPA Office of Water Regulations and Standards. *Quality Criteria for Water 1986.* Washington, DC: Environmental Protection Agency, May 1986. EPA /440/5-86-001. (May 1, 1987, update). EPA Office of Emergency and Remedial Response and Office of Waste Programs Enforcement. *Data Quality Objectives for Remedial Response Activities, Development Process.* Washington, DC: Environmental Protection Agency, March 1987. EPA/540/G-87/003.

When assessing a hazardous waste site, investigators need valid data characterizing the site, the contaminant source, and the contaminated media that could lead to human exposure. To determine if these needs are met, the investigator needs to appraise:

- health assessment data requirements,
- field data quality,
- laboratory data quality, and
- specific media considerations.

Evaluation of these criteria may reveal data irregularities or insufficiencies that may affect the health assessment. The first criterion, health assessment data requirements, is the basis for determining whether sufficient information is available for conducting a health assessment.

C.1. REVIEW DATA REQUIREMENTS

Before data can be evaluated, there must be an understanding of necessary information and criteria enabling judgment of data acceptability. The investigator should be familiar with:

- data quality objectives (the anticipated use for which samples were taken and which consequently determines types of laboratory analysis and data quality), and

- quality assurance/quality control requirements (type of information required to determine data accuracy and precision of data collected).

C.1.1. Data Quality Objectives

Data quality objectives (DQOs) are requirements needed to support decisions relating to the various stages of remedial actions (such as a health assessment). Throughout the project planning process, DQOs are supplied through qualitative and quantitative statements in documents such as sampling plans, work plans, and quality assurance plans. Familiarity with a site's DQOs will aid the data reviewer in understanding the quality of data provided and their potential limitations. In general, DQOs define data quality criteria and other analytic considerations, such as:

- definitions of criteria most commonly used to specify project data requirements and to evaluate available analytical options, namely:

  - **precision** (the reproducibility of measurements under a given set of conditions),

  - **accuracy** (the bias in a measurement system. Sources of error are the sampling process, field contamination, preservation, handling, sample matrix, sample preparation, and analysis technique.),

  - **representativeness** (the degree to which sample data accurately and precisely represent an environmental condition. This criterion is best satisfied by making certain that sampling locations are selected properly and a sufficient number of samples collected. This requires that sampling techniques and rationale for selection of sampling locations be available for the data user. Standard operating procedures (SOPs) may ensure representativeness of sampling techniques. The SOPs for groundwater sampling require that a well be purged a certain number of well volumes before sampling to assure that the sample is representative of the underlying aquifer at a point in time.),

o **completeness** (the percentage of measurements made that are judged to be valid measurements. The completeness goal is essentially the same for all data uses: a sufficient amount of valid data must be generated.),

o **comparability** (a qualitative parameter expressing the confidence with which one data set can be compared with another. This goal is achieved by using standard techniques to collect and analyze representative samples and by reporting analytical results in appropriate units.);

• guidelines and definitions for minimal quality assurance and quality control (QA/QC) sampling: by type of media, for collected samples, for duplicated samples, field blanks, background samples, and interlaboratory split samples;

• definitions of collocated, replicated, and split samples and instructions regarding applicable measurement systems, including details on sample acquisition, homogeneity, handling, shipping, storage, preparation, and analysis;

• the need for the quantity and validity of data to meet enforcement and cost recovery actions;

• reviews of internal laboratory QA/QC, including surrogate and matrix spikes, method blanks, and duplicate or replicate runs, keeping the level of required analytical support in perspective;

• reviews of other parameters, such as the effects of media variability, method detection limits, and definitions of data qualifiers and details on their interpretation and use.

DQOs are established by the Remedial Project Manager (RPM) according to anticipated data use throughout various phases of the Remedial Investigation. Five general levels of analytical options to support data collection are identified by CERCLA and described in *Data Quality Objectives for Remedial Response Activities Development Process* (1). The levels are based on the type of site to be investigated, the level of accuracy and precision required, and the intended use of the data (Table C.1.).

Level I - Field screening. This level is characterized by the use of portable instruments that can provide real-time data to assist in optimizing sampling point locations and for health and safety support. Data can be generated regarding the presence or absence of certain contaminants (especially volatile organic compounds) at sampling locations.

Level II - Field analysis. This level is characterized by the use of portable analytical instruments that can be used on-site or in mobile laboratories stationed near a site (close-support labs). Depending on the types of contaminants, sample matrix, and personnel skills, qualitative and quantitative data can be obtained.

Level III - Laboratory analysis using methods other than the Contract Laboratory Program (CLP) Routine Analytical Services (RAS). This level is used primarily in support of engineering studies using standard EPA-approved procedures. Some procedures may be equivalent to CLP RAS, without the CLP requirements for documentation.

Level IV - CLP RAS. This level is characterized by rigorous QA/QC protocols and documentation, and it

Table C.1. Summary of Analytical Levels for Data Uses

DQO Level	Data Use	Type of Analysis	Limitations	Data Quality	Cost	Time
Level I	Site characterization Monitoring during implementation	Total organic/inorganic vapor detection using portable instruments Field test kits	Instruments respond to naturally occurring compounds	If instruments calibrated and data interpreted correctly, can provide indication of contamination	Negligible excluding capital costs	Real-time
Level II	Site characterization Evaluation of alternatives Engineering design Monitoring during implementation	Variety of organics by GC; inorganics by AA; XRF Tentative identification; analyte-specific Detection limits vary from low ppm to ppb	Tentative identification Techniques/instruments limited mostly to volatiles, metals	Depends on QA/QC steps employed Data typically reported in concentration ranges	$15-$40/sample	Real-time to several hours
Level III	Risk assessment Site characterization Evaluation of alternatives Engineering design Monitoring during implementation	Organics/inorganics using EPA procedures other than CLP can be analyte-specific RCRA characteristic tests	Tentative identification in some cases Can provide data of same quality as Levels IV, NS	Similar detection limits to CLP Less rigorous QA/QC than Level IV	$960/sample for organics $200/sample for metals	14 days, but can vary based on contract requirements
Level IV	Risk assessment Evaluation of alternatives Engineering design	HSL organics/inorganics by GC/MS; AA; ICP Low ppb detection limit	Tentative identification of non-HSL parameters Some time may be required for validation of packages	Goal is data of known quality	$1,000/sample for organics $200/sample for metals	Contractually, 30-40 days Shorter turnaround time possible through SAS request
Level V	Risk assessment	Nonconventional parameters Method-specific detection limits Modification of existing methods Appendix 8 parameters	May require method development/modification Mechanism to obtain services requires special lead time	Method-specific	Initially high, if method development is required	Entries refer to all types of analysis listed. No time/cost requirements can be specified. In general, the time can range from a few weeks to significantly longer if method development is needed.

Source: (16)

provides qualitative and quantitative analytical data. Analysis may be performed by EPA regional laboratories, university laboratories, or other commercial laboratories.

Level V - Nonstandard methods. Analyses which may require method modification and/or development. CLP Special Analytical Services (SAS) are considered Level V.

As indicated previously, data quality will vary depending on anticipated data use, and specific data may undergo a variety of uses. DQOs affect factors such as detection limits, sampling locations, and analyte selection. QA/QC DQOs are of specific interest during the data review process.

C.1.2. Quality Assurance/ Quality Control

Data evaluated and used to make health assessment determinations for hazardous waste sites must meet QA/QC criteria. Quality assurance (QA) programs assure the reliability and accuracy of data. The Environmental Protection Agency (EPA) requires the development of a QA plan for all tasks involving environmental measurements (2). EPA contractors providing Remedial Investigations (RI), Feasibility Studies (FS), and related appendices are directed by EPA to use DQOs to obtain quality data. Laboratories providing analytical data in the CLP have QA/QC objectives and procedures specified in their contracts. *It is important to note that data may be generated using CLP analytical techniques, without being CLP data.* Any laboratory may use CLP analytical techniques, but "CLP data" have been reviewed by an agency unconnected to the laboratory performing the analysis, providing independent quality control information about the data quality and techniques used.

Quality control (QC), an integral part of a QA plan, is the aggregate of activities and procedures designed to ensure attainment of

prescribed quality standards for monitoring and measurement data. Quality control methods include the following: blind and identified standards analysis; multiple samples (a general term referring to any of the following types of samples: sample blanks, duplicate samples, split samples, and spiked samples); calibration of analytical equipment; and statistical design and evaluation (3,4).

The health assessor should expect both a case narrative and a data review summary to accompany EPA CLP data reviewed for a site; however, the narrative and review have rarely been included in the RI appendices for sites previously visited. Even so, EPA requires the contract laboratory to prepare a **case narrative** to document the degree to which the data conform to the data quality objectives. The case narrative contains a summary of any quality control, sample, shipment, or analytical problems. It also documents the laboratory's final solution and the internal decision process that was used. A **data review summary**, prepared by the EPA Regional Laboratory staff, documents the validation of sample holding time, instrument performance, calibration, blanks, surrogate recovery, matrix spike recovery, and compound identification. It includes documentation of actions taken to resolve data quality problems and an overall assessment. Equivalent information should be provided for non-CLP data.

When those documents are not available, the investigator should assume the data may not meet QA/QC criteria. Health assessments based on data that do not meet QA/QC criteria should include a disclaimer that acknowledges the use of possibly unreliable data and, therefore, the possibility of reaching inaccurate conclusions in the health assessment. Once familiar with data requirements for performing the health assessment, the investigator needs to review field and laboratory data.

C.2. REVIEW FIELD DATA

The investigator should be familiar with field sampling activities that may influence data acceptability. As discussed in the previous section, a case narrative should accompany the data for review. The narrative should indicate methods and strategies for sample collection, where problems may have occurred, and decisions that were made during field operations. Sample collection methods fall into two broad categories: statistical and SOPs. The statistical considerations relate to the representativeness of the data and the level of confidence that may be placed in conclusions drawn from the data. SOPs are developed to ensure sample integrity and data comparability, and to reduce sampling and analytical error. Typical sampling issues to consider:

- sample types,
- sampling strategies,
- decontamination procedures,
- sampling locations and depths,
- number of samples taken,
- sample frequency and duration,
- field quality control samples, and
- sample handling and preservation techniques.

Those topics should have been addressed in the case narrative. However, when the narrative is not present, the investigator should attempt to determine the appropriateness and acceptability of those issues from available documents. If limited or no information exists on sample collection, preservation techniques, or holding times, the data should be interpreted with caution.

C.2.1. Sample Types

When reviewing environmental data, the investigator should be familiar with, and should identify, the types of samples collected. Sample type will affect data usability and recommendations that can be substantiated. The investigator should be familiar with the following issues:

- media vs. waste samples,
- grab vs. composite samples,
- filtered vs. unfiltered samples, and
- biased vs. unbiased samples.

Media vs. waste samples. Media or environmental samples refer to sampling of air, water, soils, and other environmental media to determine the extent of contamination. Waste samples refer to the sampling of actual wastes. Typically this includes drums, impoundments, tanks, waste pipes, or other waste disposal areas. Media samples are often used to determine what compounds are present and whether those waste sources exceed any criteria or standards.

Grab vs. composite samples. Two types of samples that may be taken from environmental media are grab samples and composite samples. A grab sample contains a representative portion of a medium at a specific location at a given point in time, the representativeness of which depends on the nature of the medium sampled. A composite sample is composed of a mixture of grab samples collected at different times or locations within a medium, and it gives average contaminant concentration values. The use of composite samples can dilute isolated concentrations of hazardous waste compounds to a point below analytical detection limits, and may mask particularly high concentrations of contaminants (5) that could be detected with grab samples. For that reason, grab samples are generally preferred over composite samples at hazardous waste sites.

Filtered vs. unfiltered samples. Samples may be filtered when high levels of sediment are present. Filtering may affect the sample contaminant load.

Biased vs. unbiased samples. "Biased sampling" refers to a sampling scheme in which resulting data place emphasis on a single characteristic or factor of the problem. Unbiased sampling refers to sampling methods that allow for estimates to be drawn from the data which are representative of the receptor (i.e., potentially exposed) population or the control (unexposed) population, depending on the purpose of the sampling.

After identifying sample types that have been collected, the investigator should identify sampling strategies used to collect the data.

C.2.2. Sampling Strategies

Sampling strategy refers to the statistical technique used to locate sampling sites. Five different sampling strategies may be employed: random, systematic, stratified, judgment, and hybrid strategies (5,6,7).

Random sampling uses the theory of random chance probabilities to choose representative sample locations and, thus, no bias is involved (8). This sampling strategy is generally used when little information exists concerning the site, and it is most effective when a large number of sampling locations are used.

Systematic sampling is the strategy used most often and involves the collection of samples at predetermined, regular intervals (e.g., sampling within a grid). There is a potential for bias in this sampling strategy; for example, when the sampling sites are partially phased with regular variations of contaminant concentrations present within a medium.

Stratified sampling involves dividing the whole population of sampling sites into groups based on knowledge of the media and site characteristics. The purpose of this approach is to reduce the number of samples necessary to obtain a specified precision (4) or to increase the precision of sampling values. Precision will increase if the sample sites within the divisions are more homogeneous than those of the total population of sample sites.

Judgment sampling involves the use of human judgment in choosing sampling locations. This sampling strategy allows for investigator bias and may lead to poor quality data and improper conclusions.

Hybrid sampling is a combination of the sampling strategies just described. This sampling strategy is usually the method of choice because it allows sampling from a diverse population, and precision may be increased over other sampling strategies.

The investigator should attempt to correlate sampling strategies with data under review. After identifying sampling strategies, the investigator should identify sampling locations and the depths at which samples were collected.

C.2.3. Decontamination Procedures

Appropriate decontamination procedures should be followed to prevent contamination of samples. Field blanks (see Subsection C.2.7.) are used to determine if contamination occurred during decontamination.

C.2.4. Sampling Locations and Depths

Although sampling locations are chosen according to a sampling strategy, specific positions within the sample locations should be chosen so that samples taken from them are both representative and comparable. Because the proximity of air sampling sites to buildings, or the depth at which water samples are taken (9), can affect sample concentrations of contaminants, exact sampling sites should be based on the medium sampled and on the topographic characteristics of the hazardous waste site. The sampling locations should be accurately noted on an area map to allow for evaluation of the quality of sample location data. After identifying sampling locations, the investigator should consider the number of samples collected.

C.2.5. Number of Samples

The number of samples required to give a particular degree of precision can be statistically computed (5,10) and depends on the sampling strategy used (4). Because the environmental data may have been collected for cleanup, rather than health assessment purposes, determine if sufficient information exists for assessing the precision of the data and the variation term used in estimating precision. Preliminary information characterizing the variation within the sample is required to calculate the necessary number of samples with some degree of confidence. Equations for this computation can be found in *Characterization of Hazardous Waste Sites - A Methods Manual. Volume 1. Site Investigations* (7) and *Preparation of Soil Sampling Protocol. Techniques and Strategies* (4). The number of samples collected is often limited by the availability of funds for sampling and analysis (7) and can have an effect on the representativeness of the samples.

The investigator should also consider sampling frequency and duration. If additional samples are considered necessary for assessment of health conditions attributable to the site, a recommendation should be made in the health assessment.

C.2.6. Sampling Frequency and Duration

The frequency and duration of environmental sampling necessary to sufficiently characterize contaminant concentrations at a hazardous waste site depend on the site environment and the medium being sampled. Frequency and duration of sampling efforts should reflect seasonal and meteorologic variations of the site. Some situations requiring frequent sampling are surface water samples, crops samples obtained during the growing season, and samples taken in areas influenced by tides (7). In general, the more rapid the change in contaminant concentration, the more frequent the sampling effort should be, at least for soil sampling (4).

The investigator may identify field quality control samples while reviewing field sampling information. Those samples will be used to determine if acceptable sampling techniques were used.

C.2.7. Field Quality Control Samples

Field quality control samples are taken to determine if contamination of samples has occurred in the field, and, if possible, to quantify the extent of contamination so that data are not lost. Regardless of the level of QC implemented, the types of QC samples are the same. Definitions and use of common QC samples follow.

Trip blanks are samples that originate from analyte-free water taken from the laboratory to the sampling site and returned to the laboratory with volatile organic samples. One trip blank accompanies each cooler containing volatile organic compounds. These samples are used to determine cross-contamination of volatile samples within the cooler.

Equipment rinsates are the final analyte-free water rinses from equipment cleaning and are collected daily during a sampling event. If analytes pertinent to the project are found in the rinsate, then they will be used to flag or assess the levels of the analyte in the samples. This comparison is made during data validation, and corrections have been made to tabulated data presented to the data user.

Field blanks refer to the water used for decontamination and steam cleaning. These blanks are used to determine if sample contamination may occur from water used during decontamination.

Field duplicates/splits for soil samples are collected, homogenized, and then split. All samples except volatile organic compounds are homogenized and split. Samples for volatile compounds should not be mixed, but selected segments of soil are taken from the length of the core and placed in glass vials.

Referee duplicates are sent to the referee QA laboratory if regulators (state or region) collect samples or if a special problem occurs during sample collection or analysis.

These samples are only of benefit if the data have undergone a QC review. This review is not normally done by the laboratory performing the laboratory analysis, but by an external agency or individual responsible for quality control. Sample handling and preservation also need to be checked as part of the quality control process.

C.2.8. Sample Handling and Preservation

The manner in which environmental samples are preserved and handled, before analysis is performed, may affect their representativeness of actual site conditions. Appropriate preservation techniques depend on sample type (e.g., water, soil, atmospheric) and the type of chemical determination made (6). Some chemical determinations are more likely to be affected by sample storage than others (11). For example, phenols are subject to biological and chemical oxidation in waste water samples; therefore, water samples examined for phenols should be analyzed as soon as possible after samples are collected (5). Also, water samples to be analyzed for lead, silver, cadmium, and zinc should be acidified to keep the metals from precipitating or adsorbing to the inside of the

sample containers (11). Other samples that require proper preservation and handling include samples containing organochlorides and PCBs, pesticides, and aromatic compounds (5).

Sample holding times and preservatives are used to ascertain the validity of the results from time of collection to analysis or preparation. However, holding times may vary between methods of analysis. Table C.2. specifies holding times and preservation techniques for samples collected for Contract Laboratory Protocol, and for RCRA SW-846 analytical methods. Knowledge of these methods is important when reviewing laboratory data.

C.3. REVIEW LABORATORY DATA

Generally, the investigator will not have the technical expertise to perform validation of laboratory data. This is why the case narrative and data review summary should be supplied to the investigator. Nonetheless, the investigator needs to be familiar with laboratory terms and practices that affect data usability and interpretation. The reviewer should be familiar with the following elements:

- analytical methods,

- detection limits,

- QC samples, and

- data qualifiers.

C.3.1. Analytical Methods

The quality of environmental data can be affected by the choice of the laboratory sample analysis method. For this reason, regulatory statutes often designate acceptable analytical methods for specified lists of chemicals. Examples of such statutes are shown in Table C.3. Chemicals specified for testing by those regulatory statutes are shown in Appendix B.12.

Table C.2. Sample Preservatives and Holding Times

| Parameter | Contract Laboratory Protocol | | | SW-846 | | |
	Preservative	Holding time Soil	Water	Preservative	Holding time Soil	Water
VOCs by gas chromatography /mass spectrometry (GC/MS)	Cool, 4°C	10 days	10 days	Cool, 4°C	14 days	14 days
Polychlorinated biphenyls (PCBs)/ pesticides	Cool, 4°C	Extract within 10 days, analyze 40 days	Extract within 5 days, analyze 40 days	Cool, 4°C	Extract within 7 days, analyze 40 days	Extract within 7 days, analyze 40 days
Extractable organic compounds	Cool, 4°C	Extract within 10 days, analyze 40 days	Extract within 5 days, analyze 40 days	Cool, 4°C	Extract within 7 days, analyze 40 days	Extract within 7 days, analyze 40 days
Metals	HNO_3 to pH< 2	6 months	6 months	HNO_3 to pH< 2	6 months	6 months
Mercury	HNO_3 to pH< 2	26 days	26 days	HNO_3 to pH< 2	30 days	30 days
Cyanide	NaOH to pH> 12 Cool, 4°C, add 0.6 g ascorbic acid if residual chlorine present	14 days	14 days	NaOH to pH> 12 Cool, 4°C, add 0.6 g ascorbic acid if residual chlorine present	14 days	14 days
Chromium VI	HNO_3 to pH< 2	24 h	24 h	HNO_3 to pH< 2	24 h	24 h

Statute	Chemical List	Method
National Pollutant Discharge Elimination System (NPDES)	Priority Pollutants	40 CFR 136
RCRA	Appendix IX	SW-846
CERCLA	Superfund Hazardous Substances List (HSL), Target Compound List (TCL), and Target Analyte List (TAL)	CLP for water, soil and sediment

Table C.3. Regulatory Statutes and Analytical Methods

Chemical Class Analytical Detector	Analytical 40 CFR	SW-846	Method Detection Limit
Purgable Halocarbons	601	8010	0.5 ug/l
Aromatic Halocarbons	602	8020	0.5 ug/l
Volatile by GC/MS	624	8240	5-10 ug/l
Metals			
Inductively Coupled Plasma (ICP)	200.7	6010	varies/ element
Graphite Furnace Atomic Absorption (GFAA)	200 Series	7000 Series	Factor of 10-100 Lower than ICP
Lead ICP			50 ppb
GFAA			5 ppb

Table C.4. Method Detection Limits

When preparing DQOs for laboratory analysis, the project manager specifies the list (or sublist) of chemicals that the laboratory needs to analyze for. The investigator should be familiar with the reason for selecting the chemicals represented by the data because the selected chemicals may change between phases of the Remedial Investigation.

Even though recent environmental data are more likely to be available for review, every effort should be made to obtain historical data. The analytical methods, detection limits, QC samples, and data qualifiers of the historical data should be evaluated. If any of this information is missing, the information gaps should be noted.

In addition to specifying the chemicals to be analyzed for, the project manager will specify the analytical methods to be used by the laboratory. For the investigator performing a health assessment, this is important because the analytical method affects the detection limit.

C.3.2. Detection Limits

Detection limits are the lowest quantifiable concentrations that can be determined and are affected by several factors, such as: analytical instrument, analytical method, sample matrix, and laboratory procedures. The investigator

should be familiar with the three types of CLP detection limits.

Instrument Detection Limit is the lowest concentration of analyte that can be accurately determined when a compound is directly input to the instrument. Typically, this is the concentration at which the signal to noise ratio is three.

Method Detection Limit is the lowest concentration of analyte that can be accurately determined when the sample has been carried through the entire analytical protocol. This is also known as the Contract Required Detection Limit (CRDL) or Contract Required Quantitation Limit (CRQL). Table C.4. shows method differences and similarities for several "40 CFR" and "SW-846" analytical methods, and Table C.5. lists Superfund Target Compound List (TCL) CRQLs.

Reporting Limit is the detection limit that accounts for dilutions of sample, matrix interferences, sample preservation, and analysis. This is also known as the Practical Quantitation Limit (PQL).

The reporting limit is of most importance to the investigator performing a health assessment and should be recorded in the data for every analyte concentration level. The investigator

Table C.5. Target Compound List CRQL*

Chemical	Quantitation Limit** Water μg/L	Low Soil/ Sediment μg/kg	Chemical	Quantitation Limit** Water μg/L	Low Soil/ Sediment μg/kg	Chemical	Quantitation Limit** Water μg /L	Low Soil/ Sediment μg/kg
Volatile Compounds[a]			4-Methylphenol	10	330	Anthracene	10	330
Chloromethane	10	10	N-Nitroso-di-n-dipropylamine					
Bromomethane	10	10		10	330	Di-n-butylphthalate	10	330
Vinyl Chloride	10	10	Hexachloroethane	10	330	Fluoranthene	10	330
Chloroethane	10	10	Nitrobenzene	10	330	Pyrene	10	330
Methylene Chloride	5	5				Bultylbenzylphthalate	10	330
			Isophorone	10	330	3,3'-Dichlorobenzidine	20	660
Acetone	10	10	2-Nitrophenol	10	330			
Carbon Disulfide	5	5	2,4-Dimethylphenol	10	330	Benzo(a)anthracene	10	330
1,1-Dichloroethene	5	5	Benzoic acid	50	1600	Chrysene	10	330
1,1-Dichloroethane	5	5	bis(2-Chloroethoxy)methane			bis(2-Ethylhexyl)phthalate	10	330
1,2-Dichloroethane (total)	5	5		10	330	Di-n-octylphthalate	10	330
						Benzo(b)fluoranthene	10	330
Chloroform	5	5	2,4-Dichlorophenol	10	330			
1,2-Dichloroethane	5	5	1,2,4-Trichlorobenzene	10	330	Benzo(k)fluoranthene	10	330
2-Butanone	10	10	Naphthalene	10	330	Benzo(a)pyrene	10	330
1,1,1-Trichloroethane	5	5	4-Chloroaniline	10	330	Indeno(1,2,3-cd)pyrene	10	330
Carbon Tetrachloride	5	5	Hexachlorobutadiene	10	330	Dibenz(a,h)anthracene	10	330
						Benzo(g,h,i)perylene	10	330
Vinyl Acetate	10	10	4-Chloro-3-methylphenol (para-					
Bromodichloromethane	5	5	chloro-meta-cresol)	10	330	Pesticides[c]		
1,2-Dichloropropane	5	5	2-Methylnaphthalene	10	330	alpha-BHC	0.05	8
cis-1,3-Dichloropropene	5	5	Hexachloro-cyclopentadiene			beta-BHC	0.05	8
Trichloroethene	5	5		10	330	delta-BHC	0.05	8
			2,4,6-Trichlorophenol	10	330	gamma-BHC (Lindane)	0.05	8
Dibromochloromethane	5	5	2,4,5-Trichlorophenol	50	1600	Heptachlor	0.05	8
1,1,2-Trichloroethane	5	5						
Benzene	5	5	2-Chloronaphthalene	10	330	Aldrin	0.05	8
trans-1,3-Dichloropropene	5	5	2-Nitroaniline	50	1600	Heptachlor epoxide	0.05	8
Bromoform	5	5	Dimethylphthalate	10	330	Endosulfan I	0.05	8
			Acenaphthylene	10	330	Dieldrin	0.10	160
4-Methyl-2-pentanene	10	10	2,6-Dinitrotoluene	10	330	4,4'-DDE	0.10	16
2-Hexanone	10	10						
Tetrachloroethene	5	5	3-Nitroaniline	50	1600	Endrin	0.10	16
Toluene	5	5	Acenaphthene	10	330	Endosulfan II	0.10	16
1,1,2,2-Tetrachloroethane	5	5	2,4-Dinitrophenol	50	1600	4,4'-DDD	0.10	16
			4-Nitrophenol	50	1600	Endosulfan sulfate	0.10	16
Chlorobenzene	5	5	Dibenzofuran	10	330	4,4'-DDT	0.10	16
Ethyl Benzene	5	5						
Styrene	5	5	2,4-Dinitrotoluene	10	330	Methoxychlor	0.5	80
Xylenes	5	5	Diethylphthalate	10	330	Endrin ketone	0.10	16
			4-Chlorophenyl-phenyl ether			alpha-Chlordane	0.5	80
Semivolatile Compounds[b]				10	330	gamma-Chlordane	0.5	80
Phenol	10	330	Fluorene	10	330	Toxaphene	1.0	160
bis(2-Chloroethyl) ether	10	330	4-Nitroaniline	50	1600			
2-Chlorophenol	10	330				Aroclor-1016	0.5	80
1,3-Dichlorobenzene	10	330	4,6-Dinitro-2-			Aroclor-1221	0.5	80
1,4-Dichlorobenzene	10	330	methylphenol	50	1600	Aroclor-1232	0.5	80
			N-nitrosodiphenylamine	10	330	Aroclor-1242	0.5	80
Benzyl alcohol	10	330	4-Bromophenyl-phenylether			Aroclor-1248	0.5	80
1,2-Dichlorobenzene	10	330		10	330			
2-Methylphenol	10	330	Hexachlorobenzene	10	330	Aroclor-1254	1.0	160
bis(2-Chlorisopropyl)ether			Pentachlorophenol	50	1600	Aroclor-1260	1.0	160
	10	330	Phenanthrene	10	330			

[a]Medium Soil/Sediment CRQL for Volatile TCL compounds are 125 times the individual Low Soil/Sediment CRQL.
[b]Medium Soil/Sediment CRQL for Semivolatile TCL compounds are 60 times the individual Low Soil/Sediment CRQL.
[c]Medium Soil/Sediment CRQL for Pesticides/PCB TCL compounds are 15 times the individual Low Soil/Sediment CRQL.
*Specific quantitation limits are highly matrix dependent. The quantitation limits listed herein are provided for guidance and may not always be achievable.

should note if the PQL for an analyte exceeds concentration levels for which there may be health concerns. Problems such as this should be addressed in the health assessment.

C.3.3. Data Validation

Data validation identifies invalid data and qualifies the usability of the remaining data. This should only be done by trained individuals with appropriate technical expertise. Nonetheless, the investigator should be familiar with validation practices and their impact on the data.

Data validation is a series of performance checks and decisions on analytical procedures, calibration, instrument detection limits, QC samples, out-of-control events and corrective actions, and data evaluation and reduction. Considerations in determining the validity of the data from collected samples include the specificity and sensitivity, accuracy, and precision of the analytical procedure employed.

Specificity and sensitivity. Each analytical method used to determine the quantity of a compound in a sample has its own specificity and sensitivity. The degree of specificity and sensitivity of the method employed affects data quality. Specificity describes the ability of an analytical test to accurately detect the compound of interest. For example, an analytical test that measures the concentration of organochloride compounds in a sample may have less ability to detect one type of organochloride (e.g., 2,4-D) than a test designed to specifically measure that one type of organochloride. The sensitivity of an analytical method refers to its ability to detect small amounts of contaminant in a sample. The sensitivity of an analytical test is indicated by its detection limits or level of detection. The lower the detection limits, the greater the sensitivity. Because analytical methods can only detect the presence of a compound at concentrations at and above the level of detection, the smaller the quantity of contaminant in a

sample, the more sensitive the method must be.

Accuracy. Accuracy is a measure of the bias in the experimental system, or the measure of closeness between a resulting value (or average of a set of measurements) and the true value (known or standard value). Accuracy is expressed as absolute error, relative error, or bias (2,12). The more accurate the measurement data, the better the quality of the data.

Precision. Precision is the agreement or reproducibility of a set of repeated measurements made under the same measurement system. Precision is expressed as the range of measurement values obtained, or in terms of the standard deviation or variance of the sample measurements (12). When precision is expressed in terms of multiples of the standard deviation, it is reported (in the same units as the observed measurement) as a plus and minus range around the reported value (8).

The data user must keep the level of concern and the end use of the data in mind when reviewing precision and accuracy information. In some cases, even data of poor precision and/or accuracy may be useful. For example, if all the results are far above the level of concern, the precision and accuracy are much less important. However, when results are close to the level of concern, precision and accuracy are quite important and should be carefully reviewed. If results have very good precision but poor accuracy, it may be acceptable to correct the reported results using the percent recovery or percent bias data.

Validation of these parameters occurs through the analysis of laboratory QC samples. Common QC samples that are used for this process are:

Method Blank or Preparation Blank. This is analyte-free water or sand processed in conjunction with samples, using all the reagents, surrogates, and internal standards used in processing samples.

Matrix Spike. This is an aliquot of sample that is spiked with known concentrations of compounds of interest and subjected to the entire analytical procedure. The results of the recovery of these compounds are used to indicate the appropriateness of the method for the sample matrix.

Matrix Spike Duplicate. This is a second aliquot of sample that is spiked in the same manner as the matrix spike. Data from matrix spikes and spike duplicates are evaluated individually for recovery and percent difference. These are used with organic compounds.

Duplicate. A sample aliquot is taken and analyzed, and data are compared with the original sample. Used for wet chemical and metals analysis.

Surrogates. Compounds of similar chemical composition, spiked into each volatile and semivolatile sample before sample preparation. Recoveries are monitored.

Data qualifiers are produced during the data validation process and are listed with the analyte concentration in the data report. These qualifiers are used to classify sample data as to their conformity to QC requirements. Several data qualifiers are listed in Table C.6. The most common data qualifiers are:

- J Estimate, qualitatively correct but quantitatively suspect.

- R Reject, data not suitable for any purpose.

- U Not detected at a specified detection limit (e.g., 10U).

Sample data can be qualified with a "J" or "R" for many different reasons. Poor surrogate recovery, blank contamination, or calibration problems, among other factors, can cause sample data to be qualified. Whenever sample data are qualified, the reasons for the qualification are stated in the data validation report. The investigator should note that data validation is performed using strict analytical criteria that do not take the sampling activity's DQOs into account.

C.4. REVIEW MEDIA-SPECIFIC DATA

As part of the data evaluation process, the investigator needs to consider media-specific factors that may influence data analysis and interpretation. These factors include sampling and analytical considerations and are important for accurate interpretation of environmental and exposure pathways. The kinds of information needed to evaluate the pathways for each medium are detailed below.

C.4.1. Water

Water sampling procedures should be designed to characterize and delineate the extent and concentrations of contaminant migration. Representative sampling of both surface waters and groundwaters upgradient and downgradient of the site is necessary to distinguish health implications associated with the site from possible water contamination that is not site-related.

Surface water and groundwater contamination levels, as well as contaminant plume definition, are important and necessary components of the evaluation of groundwater contamination. An

Table C.6. Data Qualifiers

Organic Data Qualifiers

U Indicates compound was analyzed for but not detected. The sample quantitation limit must be corrected for dilution and for percent moisture.

J Indicates an estimated value. This flag is used either when estimating a concentration for tentatively identified compounds (TICs) where a 1:1 response is assumed, or when the mass spectral data indicate the presence of a compound that meets the identification criteria but the result is less than the sample quantitation limit but greater than zero.

C This flag applies to pesticide results when the identification has been confirmed by gas chromatography/mass spectrometry. Single component pesticides = 10 ng/μL in the final extract shall be confirmed by gas chromatography/mass spectrometry.

B This flag is used when the analyte is found in the associated blank as well as in the sample. It indicates possible/probable blank contamination and warns the data user to take appropriate action. This flag must be used for a TIC as well as for a positively identified TCL compound.

E This flag identifies compounds whose concentrations exceed the calibration range of the gas chromatography/mass spectrometry instrument for that specific analysis. This flag will not apply to pesticides/PCBs analyzed by GC/EC methods. If one or more compounds have a response greater than full scale, the sample or extract must be diluted and reanalyzed. If the dilution of extract causes any compounds identified in the first analysis to be below the calibration range in the second analysis, then the results of both analyses shall be reported.

D This flag identifies all compounds identified in an analysis at a secondary dilution factor. If a sample or extract is reanalyzed at a higher dilution factor, as in the "E" flag above, the "DL" suffix is appended to the sample number for the diluted sample, and all concentration values reported on that sample are flagged with the "D" flag.

A This flag indicates that a TIC is a suspected aldol-condensation product.

X Other specific flags and footnotes may be required to properly define the results. If used, they must be fully described and such description attached to the Sample Data Summary Package and the Case Narrative. If more than one is required, use "Y" and "Z" as needed. If more than five qualifiers are required for a sample result, use the "X" flag to combine several flags, as needed. For instance, the "X" flag might combine the "A," "B," and "D" flags for some samples.

R Quality control indicates that data are not usable (compound may or may not be present). Resampling and reanalysis are necessary for verification.

Q No analytical result.

Inorganic Data Qualifiers

E The reported value is estimated because of the presence of interference.

M Duplicate injection precision not met.

N Spiked sample recovery not within control limits.

S The reported value was determined by the Method of Standard Addition.

W Postdigestion spike for furnace atomic absorption analysis is out of control limits (85-115%), while sample absorbance is less than 50% of spike absorbance.

* Duplicate analysis not within control limits.

+ Correlation coefficient for the Method of Standard Addition is less than 0.995.

M (Method Qualifiers)

P   ICP

A   Flame AA

F   Furnace AA

CV Manual Cold Vapor AA

AV Automated Cold Vapor AA

AS Semiautomated Spectrophotometric

C   Manual Spectrophotometric

T   Titrimetric

Source: (17, 18)

equally important, but often neglected, component of the evaluation of contaminated water is information about contaminant concentrations in water used by humans for drinking and nondrinking purposes. Groundwater contaminant concentrations are typically well-documented, but contaminant concentrations "at the tap" are not. Those concentrations are most important in determining human health exposures. When information about contaminant concentrations in water used for drinking and nondrinking purposes is not present, it should be recommended that this concentration information be obtained. It might also be useful to recommend that the information be obtained by a residential well survey conducted by a state or local health department, or by the U.S. Geological Survey (USGS).

All surface water bodies on or contiguous with the site should be sampled, including perennial and intermittent streams and seeps that could transport contaminants away from the site, and any surface water body that may have received surface- or groundwater discharges from the site. A sufficient number of samples should have been collected from both upstream and downstream stations so that statistically valid contaminant concentrations can be determined.

For surface water bodies, the duration of the sampling program and frequency of sampling are important considerations. Samples should have been drawn at a regular frequency over an extended time, and should represent both maximal and minimal flow conditions. One surface water sample is not normally representative of water quality conditions because of possible variations in concentrations of the effluent being released to the water body, the stream stage, and mixing characteristics of the receiving waters.

Groundwater sampling programs require the installation of a network of monitoring wells on the site. At least one well should be located upgradient from the hazardous waste site; other wells should penetrate the deeper aquifers when there is the possibility of

confined aquifer contamination. When possible, wells should be screened at various depths in the aquifers to determine a concentration-depth profile. If there are multiple aquifers beneath the site, samples should be taken from each of them to determine if interaquifer leakage has occurred. A sufficient number of samples should be taken in order to delineate the extent of the contamination plume and direction of its migration from the site. It is important to understand that improperly screened wells may contribute to aquifer leakage and the spread of contaminants between confined aquifers. Questions should be raised about this possibility if the assessor believes that a problem exists.

C.4.2. Soil

For meaningful results, soil sampling should follow a protocol requiring random stratified sampling. When soil sample data are evaluated, the depth at which the samples were taken should be known because soil samples taken from the first few centimeters of the soil can differ greatly in contaminant concentrations from those taken one meter below the surface. The health assessor should be aware that, in many instances, reported soil data do not indicate the depth at which the soil sample was actually taken. If no depth has been indicated for the soil data reported, the soil data should be considered unspecified. Every effort should be made to determine the depth of the soil if the soil in question may represent an exposure pathway of concern. If the information is determined to be necessary, but cannot be obtained, the health assessment should note the data gap and indicate that the data are necessary before ATSDR can complete the health assessment.

Generally, the first one to two centimeters of soil are most likely to be involved in exposure, but pollutants that have been deposited by liquid spills or by long-term deposition of water-soluble materials may be found at soil depths ranging up to several meters (4). EPA and its contractors, for example, typically report the results of soil samples taken from the first 12 to 17 centimeters as surface soil findings. Those results could be misleading. If the surface is highly contaminated and the rest of the soil has minimal contamination, the results would underestimate the significance of the exposure potential.

Sediments underlying bodies of water are sampled similarly to soils. The depth at which sediment samples were taken should have been reported because contaminant concentrations in recently deposited sediment may differ from contaminant concentrations in older sediment deposits. Sediment samples taken from water bodies should be representative of different locations and generally should not be composited.

Preparation of soil samples should also be considered. In many cases, specific treatment of soil samples (e.g., acid digestion for metals analysis) is required to recover maximal contaminant amounts from the sample. However, such procedures may not be representative of the bioavailability of the contaminant under physiologic conditions.

C.4.3. Air

At many sites, characteristics of the site may make exposure to contaminants in outdoor air of much less concern than exposure to contaminants in indoor air. Air sampling should be made as relevant to human physiological conditions as possible. For example, time-integrated samples would be preferable to "grab" samples, and breathing zone samples are preferable to those less than optimally placed.

Consideration should be given to indoor air exposure sampling for volatile organic compounds in water used by households (e.g., dishwashers, showers, baths, toilets, lawn sprinklers, and house cleaning) because they may volatilize into the air and be inhaled. Because of the wide variation in factors (e.g., building tightness, air flow, temperature) that

influence the amount of volatilization of such substances from water, calculations to estimate volatilization are generally not an adequate substitute for sampling.

When remedial efforts that may have resulted in re-entrainment or volatilization of contaminants of concern from the site have been implemented, all appropriate National Institute for Occupational Safety and Health recommendations, and recommendations for optimal dust control, should have been followed to protect workers and residents living at the periphery of the site. Real-time monitoring should have been performed at the periphery of the worksite for action levels of contaminants of concern that have been determined to be protective of the health of nearby residents. For example, in some cases, real-time monitoring at the periphery of the worksite for the Primary National Ambient Air Quality Standard for total suspended particulates may be protective of the health of nearby residents.

C.4.4. Food

The quality of food concentration data depends entirely on the method of sampling and sample preparation (13). Contaminant levels in edible plant or animal species should be measured in either of two ways: 1) in portions of the contaminated tissues that are representative of those portions that will be used as foodstuffs; or 2) in foodstuffs prepared or processed from contaminated species. Concentration measurements should be obtained from the parts of the plant or animal that are consumed because there is a differential accumulation of contaminant in organs and tissues. Nonconsumed portions of the organism may have considerably greater (or lower) concentrations of the contaminant than the portion consumed, depending on partitioning abilities of the contaminant and characteristics of the tissues and organs involved. If, for example, the contaminant of concern concentrated in fat and the fat was typically removed before eating, the actual dose of contaminant could easily be less than the concentration of contaminant detected in the organism as a whole. Thus, an overestimation of the human health hazard may result. On the other hand, there would be an underestimation of human health concern if the consumed portions of the organism had a greater concentration of pollutants than the measurement obtained by sampling the whole organism.

Cooking or other processing of the food may increase or reduce the concentration of contaminant in the prepared food. For example, the concentration of some pesticide residues on agricultural crops may increase during processing procedures (14), while concentrations of other pesticides may be reduced. Examples of the loss of contaminants because of food preparation methods include reduction of some halogenated compounds in fish fillets by cooking (15) and of pesticide residues (DDT, Carbaryl, and parathion) on spinach by washing and blanching (13).

C.5. REFERENCES

1. EPA Office of Emergency and Remedial Response and Office of Waste Programs Enforcement. *Data quality objectives for remedial response activities development process.* Washington, DC: Environmental Protection Agency, March 1987; OSWER Directive 9355.0-7B, EPA 540-G87/003.

2. Department of Health Services (DHS), State of California. *The California site mitigation decision tree manual.* May 1986.

3. Hamilton CE. Quality assurance for water laboratories. In *National conference on quality assurance of environmental measurements.* Hazardous Materials Control Research Institute Information Transfer, Inc., U.S. Environmental Protection Agency, U. S. Geological Survey, and National Bureau of Standards. Denver, CO. November 27-29, 1978.

4. EPA Environmental Monitoring Systems Laboratory, Office of Research and Development. *Preparation of soil sampling protocol. Techniques and strategies.* Las Vegas, NV: Environmental Protection Agency, May 1983; EPA 600/4-83-020.

5. EPA Office of Research and Development, Environmental Monitoring Systems Laboratory. *Characterization of hazardous waste sites - A methods manual.* Vol. 2, *Available sampling methods.* Las Vegas, NV: Environmental Protection Agency, September 1983; EPA 600/4-83-040.

6. EPA Office of Solid Waste. *Test methods for evaluating solid waste. Physical chemical methods.* 2nd ed. Washington, DC: Environmental Protection Agency, 1982; SW-846.

7. EPA Office of Research and Development, Environmental Monitoring Systems Laboratory. *Characterization of hazardous waste sites - A methods manual.* Vol. 1, *Site investigations.* Las Vegas, NV: Environmental Protection Agency, April 1985; EPA 600/4-84/075.

8. EPA Environmental Monitoring and Support Laboratory. *Environmental monitoring series. Quality assurance guidelines for biological testing.* Las Vegas, NV: Environmental Protection Agency, August 1978; EPA 600/4-78-043.

9. EPA Office of Research and Development, Hazardous Waste Engineering Research Laboratory, Office of Solid Waste and Emergency Response, Office of Emergency and Remedial Response, and Office of Waste Programs Enforcement. *Guidance on remedial investigations under CERCLA.* Washington, DC: Environmental Protection Agency, June 1985; EPA 540/G-95/002.

10. EPA Environmental Monitoring Systems Laboratory, Office of Research and Development. *Soil sampling quality assurance user's guide.* Las Vegas, NV: Environmental Protection Agency, March 1989; EPA 600/8-89/046.

11. American Public Health Association, American Water Works Association, and Water Pollution Control Federation. *Standard methods for the examination of water and waste water.* 16th ed. Greenberg AE, Trussell RR, Clesceri LS, Franson MAH, eds. Washington, DC: American Public Health Association, 1985.

12. EPA Office of Research and Development, Health Effects Research Laboratory. *Manual of analytical quality control for pesticides and related compounds in human and environmental samples. A compendium of systematic procedures designed to assist in the prevention and control of analytical problems,* by Sherma J, Watts RR, Thompson JF, eds. Research Triangle Park, NC: Environmental Protection Agency, January 1979; EPA 600/1-79-008.

13. EPA Office of Toxic Substances. *Methods for assessing exposure to chemical substances.* Vol. 8, *Methods for assessing environmental pathways of food contamination.* Washington, DC: Environmental Protection Agency, September, 1986; EPA 560/5-85-008.

14. National Research Council. *Regulating pesticides in food. The Delaney paradox.* Washington, DC: National Academy Press, 1987.

15. Agency for Toxic Substances and Disease Registry. Comments on EPA draft "Public health evaluation manual" in June 25, 1986, letter to Mr. Henry L. Longest II, Director, Office of Emergency and Remedial Response, U.S. Environmental Protection Agency, Washington, DC, from Dr. Barry L. Johnson, Associate Administrator, ATSDR, Department of Health and Human Services, Atlanta, GA. 1986.

16. EPA Hazardous Site Control Division. *Guidance for coordinating ATSDR health assessment activities with the Superfund remedial process.* Washington, DC: Environmental Protection Agency, 1987; OSWER Directive 9285.4-02.

17. EPA Hazardous Site Evaluation Division. *Laboratory data validation functional guidelines for evaluating organics analyses.* Washington, DC: Environmental Protection Agency, February 1988.

18. EPA Hazardous Site Evaluation Division. *Laboratory data validation functional guidelines for evaluating inorganics analyses.* Washington, DC: Environmental Protection Agency, February 1988.

The Environmental Media Evaluation Guides (EMEGs), presented in Appendix A, provide health assessors with a means of selecting contaminants that need to be further evaluated for their potential impact on public health. This evaluation should include a detailed analysis of site-specific exposure pathways and conditions.

EMEG values are derived using methodology that incorporates standardized exposure assumptions. At some sites, the existing conditions may result in exposures that differ from those used to derive the EMEG values. In these situations, the health assessor can use the methodology presented in this section to define site-specific exposures more accurately. These exposure doses can then be compared to the appropriate toxicity values (e.g., Minimal Risk Levels) to determine if the exposures pose a potential health hazard.

The following generic equation can be used to estimate the exposure dose resulting from contact with a contaminated medium:

$$ED = \frac{C \times IR \times EF}{BW}$$

where,

ED = exposure dose;
C = contaminant concentration;
IR = intake rate of contaminated medium
EF = exposure factor;
BW = body weight;

Some standard values that may be useful in estimating exposures are shown in Table D.1.

The above equation yields the dose of a contaminant that is ingested into the gastrointestinal tract or inhaled into the respiratory tract. However, this exposure dose may not be the same as the absorbed dose, which is the dose that is absorbed across the gastrointestinal or respiratory epithelia. For risk assessment purposes, the exposure dose is more useful than the absorbed dose since the absorbed dose is seldom known for either humans or for the animal studies used for comparison purposes. The exposure dose in humans is comparable to the administered dose used in experimental animal studies to derive dose-response relationships.

Some exposure may occur on an intermittent or irregular basis. For these kinds of exposures, an exposure factor (EF) can be calculated to average-out the dose over the exposure interval. The exposure factor is calculated by multiplying the exposure frequency by the exposure duration, and dividing by the time period over which the dose is to be averaged. For example, if a child comes into contact with contaminated soil twice a week over a five-year period, the exposure factor would be:

$$EF = \frac{(2 \text{ days/week}) \times (52 \text{ weeks/year}) \times (5 \text{ years})}{(5 \text{ years}) \times (365 \text{ days/year})}$$

$$EF = 0.28$$

Therefore, in this example, the dose resulting from one exposure event would be multiplied

Table D.1. Standard Values

Body Weight
70 kg - adult, average (1)
16 kg - children 1 through 6 years old,
　　　50th percentile (1)
10 kg - infant (1)

Exposure Duration
70 years - lifetime; by convention
30 years - national upper-bound time (90th
　　　percentile) at one residence (1)
9 years - national median time (50th percentile)
　　　at one residence (1)

by 0.28 to yield the average daily dose over the 5-year period.

The use of an exposure factor gives the dose averaged over the period of exposure. However, the health assessor should recognize that some health effects may not depend on the average dose, but rather on the peak dose or some other measure of the dose rate.

The following discussion provides an overview of quantitative evaluation of human exposure through the following pathways: inhalation, water ingestion, soil ingestion, food ingestion, and dermal exposure to water and soil.

D.1. INHALATION

Inhalation is an important pathway for human exposure to contaminants that exist as atmospheric gases or are adsorbed to airborne particles or fibers. Inhalation exposure to contaminants from hazardous waste sites can occur as a result of direct release of gases and particles from an on-site facility, volatilization of gases from contaminated soils or water bodies, or resuspension of dust and particles from contaminated soil surfaces.

In order to estimate an inhalation exposure dose, the ventilation rate must be determined. Ventilation rates are often expressed as a minute volume, which is the volume of air inhaled in one minute (liters/minute). Minute volumes vary little with gender before age 12. However, adolescent males have a 50% higher minute volume than their female counterparts; and adult men, under heavy exertion, have a 35-40% higher minute volume than adult females (2).

A person's level and frequency of physical activity are major factors affecting minute volume. Values for average levels of activity have been established based on eight hours per day spent at each of the following: work, rest, and light-to-moderate activity (Exhibit D.1.). Although these values are applicable to the general population, individual persons may exhibit large variations based on their levels of physical activity.

Other factors influencing minute volume include: temperature, altitude, conditions which may aggravate air quality background levels, and a person's weight, height, health, smoking status, and pulmonary disease status (2).

Exhibit D.1. illustrates how inhalation exposure doses can be estimated and provides inhalation minute volumes.

D.2. WATER INGESTION

Ingestion of contaminated water is often the most significant source of exposure to hazardous substances from a site. To estimate exposure to a contaminant from the ingestion of potable water, the contaminant concentrations in tap water samples from individual homes should preferably be analyzed. In the absence of data on individual wells, the assessor may consider using data from monitoring wells to estimate upper limits for exposures to contaminants.

The oral ingestion dose is ideally computed using site-specific information for the populations at risk (i.e., bodyweight and consumption rates). Exhibit D.2. illustrates how exposure doses via drinking water can be estimated.

D.3. SOIL INGESTION

Soil ingestion can occur by the inadvertent consumption of soil on hands or food items, mouthing of objects, or the ingestion of nonfood items (pica). All children mouth or ingest non-food items to some extent. The degree of pica behavior varies widely in the

Exhibit D.1. Estimating Inhalation Exposure Doses

Inhalation exposure doses (IDa) can be estimated as follows:

$$IDa = \frac{C \times IR \times EF}{BW}$$

where,

IDa = inhalation exposure dose (mg/kg/day);
C = contaminant concentration (mg/m^3);
IR = inhalation rate (m^3/day);
EF = exposure factor (unitless);
BW = body weight (kg).

Standard Inhalation Values

Minute Volume	Man	Woman	Child 10 yr	Infant 1 yr	Newborn
Resting (L/min)	7.5	6.0	4.8	1.5	0.5
Light activity	20.0	19.0	13.0	4.2	1.5
Liters/Day					
8hr workday	9600	9100	6240	2500(10hrs)	90(1hr)
8hr non-occupational	9600	9100	6240	....	....
8hr rest	3600	2900	2300	1300(14hrs)	690(23hrs)
24hrs (total)	2.3×10^4	2.1×10^4	1.5×10^4	0.38×10^4	0.08×10^4
Daily inhalation	23m^3	21m^3	15m^3	3.8m^3	0.8m^3

Note:
1000 liters = 1 m^3
Source (7)

Exhibit D.2. Estimating Water Ingestion Exposure Doses

Water ingestion exposure doses (IDw) can be calculated as follows:

$$IDw = \frac{C \times IR \times EF}{BW}$$

where,

IDw = ingestion exposure dose (mg/kg/day);
C   = contaminant concentration (mg/L);
IR = ingestion rate (L/day);
EF = exposure factor (unitless);
BW = body weight (kg).

Drinking Water Standard Ingestion Values

Parameter	Value	Source
Average daily water intake of an adult	2 L/day	(1)
Average daily water intake of a child	1 L/day	(8)

Example

Consider human exposure to a primary water supply that is contaminated with 350 mg/L methyl chloride. To compute an adult exposure dose, assume a body weight of 70 kg and a water ingestion rate of 2 L/day.

$$IDw = \frac{C \times IR \times EF}{BW}$$

$$IDw = \frac{367\ mg/L \times 2\ L/day \times 1}{70\ kg}$$

$$IDw = 10\ mg/kg/day$$

For children, assume an average weight of 10 kg and a water ingestion rate of 1 L/day:

$$IDw = \frac{350\ mg/L \times 1\ L/day \times 1}{10\ kg}$$

$$IDw = 35\ mg/kg/day$$

population, and is influenced by nutritional status and the quality of care and supervision. Groups that are at an increased risk for pica behavior are children aged 1-3 years old, children from families of low socioeconomic status, and children with neurologic disorders (e.g., brain-damage, epilepsy, mental retardation) (1).

For non-pica children, a soil ingestion rate of about 50 to 100 mg per day is supported by recent studies using tracer metals in soil (4,5). Soil ingestion by adults has not been well studied, but limited evidence suggests a tentative value of 50 mg/day (6). For children with obvious pica behavior, soil ingestion rates of 5-10 grams per day are possible (1).

Both use of and accessibility to the site and surrounding areas must be considered when evaluating a site's soil exposure pathways. Sites with abandoned buildings, standing water, or streams may attract children, and exposures may occur at sites near playgrounds or school yards despite fencing and other efforts to restrict access.

Workers at commercial or industrial properties could ingest contaminated soil at rates which depend upon the type of employment.

Both residential and recreational areas are likely to provide access for exposure. Contaminated soil can be brought into homes on the feet of family members and pets. Suspended soil particulates in outdoor air can also enter a house through indoor-outdoor air exchange. A young child playing on the floor will have the maximum opportunity both for ingestion and for dermal exposure to soil and dust accumulated on the floor.

Exhibit D.3. illustrates how soil ingestion exposures can be estimated, and it provides soil ingestion rates for various age groups.

D.4. FOOD INGESTION

Assessment of the human health risk from ingestion of contaminated food requires information on the quantities of contaminated foodstuffs consumed and the extent of contamination present in foodstuffs. The most reliable method of assessing the extent of human exposure to contaminants in food is direct measurement of concentrations in foodstuffs. Such measurements should be conducted on foodstuffs prepared for consumption or portions of contaminated plants and animals that are representative of those portions used as food.

If food chains appear to be a significant pathway for human exposure and the appropriate information on contaminant levels is not available, that lack of information should be explicitly identified in the health assessment and a recommendation should be made that the appropriate information be obtained.

Estimation of exposure dose through food chains requires knowledge of the consumption rate of specific food items in the human diet. Nationwide daily consumption rates by food group are presented in Appendix E.

The consumption rates of the population in the vicinity of a hazardous waste site may differ considerably from national average consumption rates. For example, regional consumption rates of fish—freshwater, saltwater, and shellfish—may vary widely from national averages. Consumption rates of subpopulations within the contaminated area may also vary significantly from the national averages. For example, consumption of fish increases with the age of the consumer (9), and fish consumption for sport fishers is likely to be higher than the average for the United States population (10). When local consumption patterns are available, and different from national averages, they should be used in calculations to determine exposure estimates.

In the case of residential soil contamination, the consumption rate of home-grown foods and

Exhibit D.3. Estimating Soil Ingestion Exposure Doses

The soil ingestion exposure dose can be estimated as follows:

$$IDs = \frac{C \times IR \times EF\ 10^{-6}}{BW}$$

where:

IDs	=	soil ingestion exposure dose (mg/kg/day);
C	=	contaminant concentration (mg/kg);
IR	=	soil ingestion rate (mg/day);
EF	=	exposure factor (unitless);
BW	=	body weight (kg).

A conversion factor of 10^{-6} kg/mg is required to convert the soil contaminant concentration (C) from mg/kg soil to mg/mg soil

Example

Consider adult ingestion of soil with a contaminant concentration of 100 mg/kg and a daily soil ingestion rate of 50 mg/day. Assume the subject is on-site five days per week, 50 weeks per year, for 30 years. First, calculate the exposure factor:

EF = exposure frequency $\times$ exposure duration $\div$ exposure time

$$EF = \frac{(5\ days/week) \times (50\ weeks/year) \times (30\ years)}{(365\ days/year) \times (70\ years)} = 0.29$$

$$IDs = \frac{C \times IR \times EF \times 10^{-6}}{BW}$$

$$IDs = \frac{100\ mg/kg \times 50\ mg/day \times 0.29 \times 10^{-6}}{70\ kg} = 2 \times 10^{-5}\ mg/kg/day$$

local wild plants is of interest. Appendix E presents consumption rates of home-grown foods as determined by the U. S. Department of Agriculture (11). These data are organized into four groups: urban, rural non-farm, rural-farm, and all households. Table E.7. in Appendix E contains the average percentage consumption of home-grown foods.

To estimate the total daily intake of a particular contaminant, daily intakes of contaminants from **all** affected foodstuffs should be considered. Exhibit D.4. illustrates how food ingestion exposure doses can be estimated.

D.5. DERMAL EXPOSURE

Dermal absorption of contaminants from soil or water is a potential pathway for human exposure to environmental contaminants. Dermal absorption depends on numerous factors including the area of exposed skin, anatomical location of exposed skin, length of contact, concentration of chemical on skin, chemical-specific permeability, medium in which the chemical is applied, and skin condition and integrity.

The area of skin that is exposed will be influenced by the activity being performed and the season of the year. Skin surface area also varies with age. EPA (1,12) provides data on skin surface areas of different parts of the body for adults and for children.

Water

Dermal absorption of contaminants in water occurs during bathing, showering, or swimming and may be a significant route of exposure. Worker exposure via this pathway will depend on the type of work performed, protective clothing worn, and the extent and length of water contact. The permeability of the skin to a chemical is influenced by the physicochemical properties of the substance, including its molecular weight (size and shape), electrostatic charge, hydrophobicity, and solubility in aqueous and lipid media. In general, chemicals that demonstrate high skin permeability are low molecular weight, non-ionized, and lipid soluble.

Chemical-specific permeability constants should be used to estimate dermal absorption of a chemical from water. Values for dermal permeability constants may vary over a large range, spanning at least five orders of magnitude (13). Dermal permeability constants are available for relatively few chemical substances (see reference 3 for a summary of reported values). Before using a dermal permeability constant, the original reference should be checked to ensure the applicability of the experimental study. In some studies, the permeability constants were determined using neat liquids or concentrated aqueous solutions. Exposure of skin to high concentrations of organic solvents can cause delipidation of the skin, which can profoundly alter the skin's permeability. Dermal permeability constants derived from animal studies may not be applicable for human assessment purposes because of substantial differences in their skin permeability.

Dermal absorption can be significantly increased by skin abrasions which remove the outer stratum corneum layer of the skin. Pathological skin conditions, such as psoriasis or eczema, can also result in increased penetration of chemical substances into the skin (14).

When the permeability constant for a chemical is known, the dermal absorption of a chemical from water can be estimated as illustrated in Exhibit D.5.

Soil

Dermal absorption of contaminants from soil or dust depends on the area of contact, the duration of contact, the chemical and physical attraction between the contaminant and the soil, and the ability of the contaminant to penetrate the skin. Chemical specific factors, such as lipophilicity, polarity, volatility, molecular weight, and solubility also affect dermal absorption.

Many organic chemicals bind to organic matter in soil, thereby decreasing their absorption by the skin. In addition, only the fraction of the contaminant that is in direct contact with the skin is amenable to absorption. Therefore, the ability of a soil contaminant to be dermally absorbed depends on the diffusion of the contaminant through the soil matrix. Experimental studies have confirmed that dermal absorption of a contaminant may be reduced when the contaminant is applied in soil as compared to direct dermal application of the compound (15).

A soil-specific factor involved in dermal absorption is adherence, the quantity (mg/cm^2) of soil on the skin. Hawley (16) reports soil adherence values of 0.5 mg/cm^2 for children and 3.5 mg/cm^2 for adults. EPA (1,3) has reported soil adherence values of 1.45 mg/cm^2 for commercial potting soil and 2.77 mg/cm^2 for kaolin clay. Data on dust adherence to skin are limited; however, Hawley's work (16) also provides a dust adherence value of 1.8 mg/cm^2 for adults. Based on this data, a soil adherence value of 2 mg/cm^2 is proposed.

To calculate the average lifetime dermal dose, divide a 70 year lifetime exposure period into the time intervals shown in Exhibit D.6. For each exposure time interval, dermal absorption is estimated as the soil concentration times the soil adhered times the fractional lifetime exposure. This product is divided by the appropriate body weight for each exposure time interval. Exhibit D.6. illustrates how soil dermal absorbed doses can be estimated.

Exhibit D.4. Estimating Food Ingestion Exposure Doses

The food ingestion dose of a contaminant can be estimated in the following manner:

$$IDf = \sum_{i=1}^{n} \frac{CL \times CR_i \times EF}{BW}$$

where,

IDf   =   Food ingested exposure dose (mg/kg/day);
CLi   =   Concentration of contaminant in food group i (mg/g);
CRi   =   Consumption rate of food group i (g/day);
EF   =   Exposure factor (unitless);
BW   =   Body weight (kg);
n   =   Total number of food groups.

The calculation of food ingestion dose of contaminant for homegrown foods is similar, but takes into account the percentage of contaminated food that is homegrown:

$$IDf = \sum_{i=1}^{n} \frac{CL_i \times CR \times EF \times PH_i}{BW}$$

where,

PHi   =   Percentage of food group that is homegrown (Table E.2., Appendix E).

Example

The following example illustrates calculation of the food ingestion exposure dose for cadmium through garden crop contamination. The symbols used are defined above. The consumption rates (CR) and percentage of food that is home-grown (PH) were obtained from Appendix E.

Food	CL	CR	PH	EF	BW	Daily Intake of Contaminant (mg/kg/day)
Potatoes	0.02	88	9.30	1	70	0.002
Dark-green vegetables	0.01	15	21.20	1	70	0.0005
Deep-yellow vegetables	0.51	15	21.20	1	70	0.02
Tomatoes	0.24	38	21.20	1	70	0.03
Other vegetables	0.01	136	21.20	1	70	0.004
Total vegetables/fruit						0.05

Thus, the daily human intake of cadmium from contaminated garden produce in this example is estimated to be 0.05 mg/kg/day. Estimates should be confirmed, as necessary, by a local consumption survey.

Exhibit D.5. Estimating Water Dermal Absorbed Doses

Water dermal-absorbed doses can be estimated using the following formula:

$$DDw = \frac{C \times P \times SA \times ET}{BW} \times \frac{1\ liter}{1000\ cm^3}$$

where:

DDw =   Dermal absorbed dose from water (mg/kg/day);
C   =   Contaminant concentration in water (mg/L);
P   =   Permeability constant (cm/hr);
SA   =   Exposed body surface area (cm^2);
ET   =   Exposure time (hours/day);
BW   =   Body weight (kg).

The term 1 liter/1,000 cm^3 is a volumetric conversion constant.

Dermal Standard Exposure Values
50th Percentile Total Body Surface Area (cm^2)

Age (yrs)	Male	Female
3 < 6	7280	7110
6 < 9	9310	9190
9 < 12	11600	11600
12 < 15	14900	14800
15 < 18	17500	16000
18 – 70	19400	16900

50th Percentile Body Part-Specific Surface Areas for Males (cm^2)

Age(yrs)	Arms	Hands	Legs
3 < 4	960	400	1800
6 < 7	1100	410	2400
9 < 10	1300	570	3100
18 – 70	2300	820	5500

Source: 1,12

Exhibit D.6. Estimating Soil Dermal Absorbed Doses

The soil dermal exposure dose can be estimated as follows:

$$DDs = \frac{C \times A \times BF \times EF \times 10^{-6}}{BW}$$

where,

DDs = dermal absorbed dose from soil (mg/kg/day);
C = contaminant concentration in soil (mg/kg);
A = total soil adhered (mg);
BF = bioavailability factor (unitless);
EF = exposure factor (unitless);
BW = body weight (kg).

A conversion factor of 10^{-6} kg/mg is used to convert the soil contaminant concentration (C) from mg/kg soil to mg/mg soil.

The total soil adhering to the dermal surface is estimated as the product of the exposed dermal area and the soil adherence concentration.

Soil Dermal Standard Exposure Values (12)

Age (years)	Body Weight (kg)	Total Surface Area (cm^2)	% Area Exposed	Exposed Area (cm^2)	Total Soil Adhered (mg)
0 - 1	10	3500	30	1050	2100
1 - 11	30	8750	30	2625	5250
12 - 17	50	15235	28	4300	8600
18 - 70	70	19400	24	4700	9400

Example

Estimate the average daily absorbed dose for a child that has been exposed to a soil contaminant at 100 mg/kg every day from birth through 11 years of age. Assume that the average exposed skin surface area during this time is 30% and the bioavailability factor for the contaminant is 0.1.

$$DDs = \frac{C \times A \times BF \times EF \times 10^{-6}}{BW} + \frac{C \times A \times BF \times EF \times 10^{-6}}{BW}$$
(exposure for age 0−1 + age 1−11)

$$DDs = \frac{100 mg/kg \times 2100 mgkg \times 0.1 \times (1/11) \times 10^{-6}}{10 kg} + \frac{100 mg/kg \times 5250 mg \times 0.1 \times (10/11) \times 10^{-6}}{30 kg}$$

DDs = 0.002 mg/kg/day

D.6. REFERENCES

1. EPA Office of Health and Environmental Assessment. *Exposure factors handbook.* Washington, DC: Environmental Protection Agency, March 1990; EPA/600/8-89/043.

2. National Council on Radiation Protection and Measurements (NCRP). *Radiological assessment: predicting the transport, bioaccumulation, and uptake by man of radionuclides released to the environment.* Bethesda, MD: National Council on Radiation Protection and Measurements, 1984. NCRP Report no. 76.

3. EPA Office of Emergency and Remedial Response, Office of Solid Waste and Emergency Response. *Superfund exposure assessment manual.* Washington, DC: Environmental Protection Agency, 1988; EPA 540/1-88/001.

4. Calabrese EJ, et al. How much soil do young children ingest: an epidemiologic study. *Regulatory Toxicology and Pharmacology* 1989; 10:123-37.

5. Davis S, et al. Quantitative estimates of soil ingestion in normal children between the ages of 2 and 7 years. *Archives of Environmental Health* 1990; 45:112-22.

6. Calabrese J, et al. Preliminary adult soil ingestion estimates: results of a pilot study. *Regulatory Toxicology and Pharmacology* 1990; 12:88-95.

7. International Commission on Radiological Protection (ICRP). *Report of the task group on reference man.* New York: Pergamon Press, 1975.

8. National Academy of Sciences (NAS). *Drinking water and health.* Washington, DC: NRC Press, 1977.

9. Rupp EM, Miller FL, Baes CF III. Some results of recent surveys of fish and shellfish consumption by age and region of U. S. residents. *Health Physics* 1980; 39(2):165-75.

10. Office of Toxic Substances, Exposure Division. *Methods for assessing exposure to chemical substances. Vol. 8, Methods for assessing environmental pathways of food contamination.* Washington, DC: Environmental Protection Agency, September 1986; EPA 560/5-85-008 and PB87-107850.

11. U.S. Environmental Protection Agency. *Dietary consumption distributions of selected food groups for the U.S. population.* Washington, DC: Environmental Protection Agency, February 1980, EPA 560/11-80-012 and PB81-147035.

12. EPA Office of Health and Environmental Assessment. *Development of statistical distributions or ranges of standard factors used in exposure assessments.* Washington, DC: Environmental Protection Agency, March 1985; OHEA-E-161.

13. Duggard PH. Absorption through the skin: theory, *in vitro* techniques, and their applications. *Food Chemistry and Toxicology* 1986; 24:749-53.

14. Hensby CN, Schaefer H, Schalla W. The bioavailability and toxicity of topical drugs related to diseased skin. In: Chambers PL, Gehring P, and Sakai F, eds. *New concepts and developments in toxicology.* New York, NY: Elsevier Science Publishers, 1986.

15. Yang JJ, *et al. In vitro* and *in vivo* percutaneous absorption of benzo[a]pyrene from petroleum crude-fortified soil in the rat. *Bulletin of Environmental Contamination and Toxicology* 1989; 43:207-14.

16. Hawley JK. Assessment of health risk from exposure to contaminated soil. *Risk Analysis* 1985; 5(4):289-302.

The following tables provide information concerning the average daily food intake per individual per day for:

Meat, Poultry, and Fish . . . . E.1.

Milk, Milk Products, Eggs, Legumes, Nuts, and Seeds . . . . . . E.2.

Grain Products, Fats, and Oils . . . . E.3.

Vegetables . . . . E.4.

Fruits . . . . E.5.

Sugar, Sweets, and Beverages . . . . E.6.

Average Percentage of Seasonal and Annual Consumption
of Various Homegrown Foods . . . . E.7.

Table E.1. Average Intake[1] Per Individual Per Day[2], of Meat, Poultry, and Fish, 1977-78, 48 States, All Urban Areas

Sex and Age (Years)	Individuals	Total	Beef	Lamb, Pork	Veal, Game	Poultry		Organ Meats, Mixtures	Frankfurters, Sausages, Luncheon Meat	Fish, Shell-Fish	Mixtures (Mainly Meat, Poultry, Fish)
						Total	Chicken				
						--- Grams ---					
Males and Females:											
Under 1	421[3]	55	6	3	2	6	5	1	2	-[4]	35
1-2	1,035[3]	103	20	8	1	14	12	1	15	4	40
3-5	1,719	122	27	11	1	16	15	1	17	6	44
6-8	1,841	154	34	13	1	20	18	1	18	7	60
Males:											
9-11	939	189	43	16	2	23	20	2	19	8	76
12-14	1,150	216	52	18	1	26	23	1	22	9	84
15-18	1,394	267	66	25	2	30	26	2	25	10	107
19-22	1,030	290	76	25	1	32	28	3	29	14	109
23-34	2,716	292	75	28	3	30	26	2	30	16	107
35-50	2,571	288	78	30	3	31	27	3	28	15	101
51-64	2,161	266	71	29	3	30	27	5	26	18	84
65-74	1,049	226	51	24	4	28	24	4	18	17	79
75 and over	465	206	54	27	4	21	19	3	16	10	70
Females:											
9-11	1,011	164	40	14	1	20	18	1	18	6	65
12-14	1,148	179	42	16	1	21	19	1	18	9	71
15-18	1,473	186	45	16	2	22	20	1	16	11	73
19-22	1,317	183	43	19	1	24	22	1	16	11	69
23-34	3,879	187	46	17	2	23	21	2	16	11	69
35-50	3,759	191	51	19	2	24	20	3	14	13	65
51-60	2,936	190	46	19	3	25	22	3	14	14	67
65-74	1,376	165	38	17	4	24	22	3	11	12	55
75 and over	751	148	34	16	3	22	20	3	12	8	51
All Individuals	36,142[3]	204	51	20	2	24	22	2	19	12	74

[1] Quantities given are for foods as ingested; no inedible parts are included.
[2] Based on 3 consecutive days of dietary intake.
[3] Excludes breast-fed infants.
[4] Value less than 0.5 but more than 0.

Note: See Appendix E Table Definitions for foodstuffs considered in each category.

Source: USDA Nationwide Food Consumption Survey, 1977-78, 48 conterminous States (1).

Table E.2. Average Daily Intake of Milk, Milk Products, Eggs, Legumes, Nuts, and Seeds[1]

| Sex and Age | Individuals | Total Calcium Equivalent[2] | Milk, Milk Products, Milk Drinks | | | Cream Milk Desserts | Cheese | Eggs | Legumes, Seeds, Nuts |
			Total	Fluid Milk	Yogurt				
			Grams						
Males and Females:									
Under 1	421[3]	495	669	342	2	8	2	5	59
1-2	1,035[3]	475	418	401	2	14	8	21	2
3-5	1,719	449	385	361	2	20	9	20	22
6-8	1,841	537	466	426	2	26	9	17	25
Males:									
9-11	939	559	487	440	1	34	7	20	28
12-14	1,150	612	533	473	1	34	9	22	34
15-18	1,394	642	538	482	2	34	13	31	32
19-22	1,030	463	368	334	2	19	15	33	27
23-34	2,716	381	265	240	4	23	20	35	29
35-50	2,571	301	201	182	3	26	18	37	31
51-64	2,161	312	203	190	2	32	19	38	27
65-74	1,049	312	220	208	2	30	17	37	23
75 and over	465	317	226	221	0	31	15	39	24
Females:									
9-11	1,011	519	449	405	1	29	9	17	28
12-14	1,148	468	394	343	2	30	10	17	24
15-18	1,473	403	329	288	3	22	11	18	21
19-22	1,317	297	221	194	5	16	14	23	23
23-34	3,879	278	191	169	5	15	18	23	20
35-50	3,759	215	142	127	4	16	17	25	19
51-60	2,936	232	154	143	4	21	18	25	18
65-74	1,376	248	171	161	5	23	17	23	14
75 and over	751	274	198	184	2	27	17	21	14
All Individuals	36,142[3]	365	283	254	3	23	15	26	24

[1] Quantities given are for foods as ingested; no inedible parts are included and numbers are based on 3 consecutive days of dietary intake.

[2] Calcium Equivalent is quantity of whole fluid milk to which dairy products (except butter) are equivalent in calcium content.

[3] Excludes breast-fed infants.

Note: See Appendix E Table Definitions for foodstuffs considered in each category.

Source: USDA Nationwide Food Consumption Survey, 1977-78, 48 conterminous States (1).

Table E.3. Average Daily Intake of Grain Products, Fats, and Oils[1]

Sex and Age (Years)	Individuals	Grain Products Total	Bread Rolls, Biscuits	Other Baked Goods	Cereals, Pastas Total	Cereals, Pastas Ready-To-Eat Cereals	Mixtures Mainly Grain	Fats and Oils Total	Table Fats	Salad Dressing
					--- Grams ---					
Males and Females:										
Under 1	421[2]	65	3	5	45	28	12	3	3[3]	3[3]
1-2	1,035[2]	161	31	27	55	13	48	5	3	1
3-5	1,719	198	47	41	56	15	55	7	5	3
6-8	1,841	227	56	51	59	18	60	9	5	3
Males:										
9-11	939	261	64	63	63	19	71	10	6	4
12-14	1,150	292	76	66	63	20	87	12	7	4
15-18	1,394	304	89	73	64	17	78	14	8	5
19-22	1,030	258	84	59	51	10	64	13	7	6
23-34	2,716	261	86	61	47	8	69	17	8	8
35-50	2,571	247	82	58	50	7	57	18	9	8
51-64	2,161	237	81	61	54	10	41	19	9	8
65-74	1,049	230	74	56	68	13	31	16	9	5
75 and over	465	242	72	63	77	12	30	15	10	3
Females:										
9-11	1,011	241	57	56	58	17	71	10	5	4
12-14	1,148	231	58	57	53	13	64	10	5	4
15-18	1,473	202	55	47	43	9	57	12	5	6
19-22	1,317	184	51	36	41	6	56	12	5	6
23-34	3,879	179	51	39	38	6	50	14	6	7
35-50	3,759	169	52	38	37	5	41	14	5	7
51-64	2,936	169	55	40	41	8	33	14	6	6
65-74	1,376	178	55	40	49	10	34	12	6	4
75 and over	751	190	55	44	63	11	28	14	8	4
All Individuals	36,142[2]	213	62	49	50	10	52	13	6	6

[1] Quantities given are for foods as ingested; no inedible parts are included; based on 3 consecutive days of dietary intake.
[2] Excludes breast-fed infants.
[3] Value less than 0.5 but more than 0.

Note: See Appendix E Table Definitions for foodstuffs considered in each category.

Source: USDA Nationwide Food Consumption Survey, 1977-78, 48 conterminous States (1).

Table E.4. Average Daily Intake of Vegetables[1,2]

Sex and Age (Years)	Individuals	Total	White Potatoes	Tomatoes	Dark-Green Vegetables	Deep-Yellow Vegetables	Other Vegetables
				--Grams--			
Males and Females:							
Under 1	421[3]	77	10	1	2	18	46
1-2	1,035[3]	98	34	10	3	5	45
3-5	1,719	110	41	13	3	5	49
6-8	1,841	145	53	13	5	6	67
Males:							
9-11	939	167	66	16	6	7	73
12-14	1,150	187	74	18	7	8	80
15-18	1,394	216	88	22	8	7	91
19-22	1,030	217	83	25	7	6	95
23-34	2,716	233	81	29	9	8	106
35-50	2,571	256	82	32	10	10	122
51-64	2,161	275	79	38	11	11	136
65-74	1,049	256	73	33	12	15	124
75 and over	465	250	74	33	15	13	115
Females:							
9-11	1,011	162	58	16	6	6	76
12-14	1,148	160	60	18	5	5	71
15-18	1,473	163	60	18	6	6	72
19-22	1,317	170	53	24	5	5	82
23-34	3,879	187	52	26	8	8	93
35-50	3,759	201	51	29	10	8	102
51-60	2,936	224	54	33	11	11	114
65-74	1,376	224	53	32	12	14	113
75 and over	751	211	53	29	12	14	102
All Individuals	36,142[3]	198	62	25	8	9	95

[1] Quantities given are for foods as ingested; no inedible parts are included; based on 3 consecutive days of dietary intake.
[2] Mixtures are included in each subgroup and in the total.
[3] Excludes breast-fed infants.

Note: See Appendix E Table Definitions for foodstuffs considered in each category.

Source: USDA Nationwide Food Consumption Survey, 1977-78, 48 conterminous States (1).

Table E.5. Average Daily Intake of Fruits[1]

Sex and Age (Years)	Individuals	Total	Citrus Fruits Total	Citrus Fruits Juices	Dried Fruits	Other Fruits, Mixtures, Juices Total	Apples	Bananas	Other Fruits, Mixtures Mainly Fruit	Noncitrus Juices, Nectars
						Grams				
Males and Females:										
Under 1	421[2]	151	21	21	3[3]	130	18	16	64	31
1-2	1,035[2]	150	60	54	1	89	21	14	27	27
3-5	1,719	135	61	53	1	73	20	9	23	21
6-8	1,841	153	68	58	1	85	26	9	34	16
Males:										
9-11	939	143	65	56	3[3]	78	25	8	35	10
12-14	1,150	141	66	55	3[3]	74	26	8	34	7
15-18	1,394	138	71	61	3[3]	68	21	7	30	9
19-22	1,030	114	60	50	3[3]	53	17	9	21	8
23-34	2,716	123	62	53	3[3]	61	16	7	26	12
35-50	2,571	132	59	48	1	73	21	8	35	8
51-64	2,161	169	73	53	1	95	23	14	49	8
65-74	1,049	182	70	53	3	109	18	15	63	13
75 and over	465	183	67	47	4	112	28	16	52	15
Females:										
9-11	1,011	155	69	59	3[3]	85	26	8	38	12
12-14	1,148	135	63	53	3[3]	72	21	7	33	10
15-18	1,473	118	60	51	3[3]	57	18	6	24	9
19-22	1,317	117	62	52	1	54	12	5	26	11
23-34	3,879	122	61	51	1	60	16	6	26	12
35-50	3,759	125	62	50	1	62	15	6	32	9
51-64	2,936	177	80	61	1	96	22	11	51	12
65-74	1,376	189	87	65	2	101	18	14	55	13
75 and over	751	181	71	52	4	107	24	15	51	16
All Individuals	36,142[2]	142	66	54	1	76	20	9	35	12

[1] Quantities given are for foods as ingested; no inedible parts are included; based on 3 consecutive days of dietary intake.
[2] Excludes breast-fed infants.
[3] Value less than 0.5 but more than 0.

Note: See Appendix E Table Definitions for foodstuffs considered in each category.

Source: USDA Nationwide Food Consumption Survey, 1977-78, 48 conterminous States (1).

Table E.6. Average Daily Intake of Sugar, Sweets, and Beverages[1]

Sex and Age (Years)	Individuals	Total	Sugar	Candy	Total	Nonalcoholic Beverages					Alcoholic Beverages	
						Total	Coffee	Tea	Soft Drinks	Fruit Drinks, Ades	Total	Beer, Ale
						— Grams —						
Males and Females:												
Under 1	421[2]	6	[3]	[3]	22	22	0	6	10	6	0	0
1-2	1,035[2]	17	2	3	153	153	1	27	93	31	[3]	[3]
3-5	1,719	24	3	3	216	216	1	39	141	34	[3]	[3]
6-8	1,841	29	3	4	232	232	2	47	146	37	[3]	0
Males:												
9-11	939	31	3	6	277	277	4	55	179	38	[3]	0
12-14	1,150	36	4	6	336	336	9	83	203	41	[3]	0
15-18	1,394	31	4	5	484	466	40	100	283	43	18	16
19-22	1,030	19	5	4	692	577	113	116	312	37	114	109
23-34	2,716	23	6	3	933	760	311	151	271	27	173	154
35-50	2,571	24	7	2	1,012	859	523	157	162	18	153	130
51-64	2,161	27	7	2	902	805	559	142	94	11	96	78
65-74	1,049	29	7	2	714	658	459	134	54	11	56	45
75 and over	465	29	7	1	623	589	426	100	42	20	34	25
Females:												
9-11	1,011	29	3	5	253	253	3	58	156	36	[3]	[3]
12-14	1,148	27	3	6	318	318	8	75	200	35	[3]	[3]
15-18	1,473	23	3	5	434	430	47	92	259	33	5	3
19-22	1,317	15	4	3	570	536	126	120	265	25	35	24
23-34	3,879	16	5	2	733	693	284	171	217	21	40	24
35-50	3,759	18	5	2	832	803	463	172	152	16	29	15
51-64	2,936	19	4	2	766	739	484	158	84	13	27	16
65-74	1,376	22	4	1	609	599	382	155	49	12	10	5
75 and over	751	22	4	1	540	531	350	140	30	11	9	5
All Individuals	36,142[2]	23	5	3	625	578	265	123	167	24	47	38

[1] Quantities given are for foods as ingested; no inedible parts are included; based on 3 consecutive days of dietary intake.
[2] Excludes breast-fed infants.
[3] Value less than 0.5 but more than 0.

Note: See Appendix E Table Definitions for foodstuffs considered in each category.

Source: USDA Nationwide Food Consumption Survey, 1977-78, 48 conterminous States (1).

Table E.7. Average Percentage of Seasonal and Annual Consumption of Homegrown Foods. (All Households)

Food	Annual	Spring	Summer	Fall	Winter
Milk, cream, cheese	4.00	3.88	3.96	4.82	3.47
Fats, oils	1.89	2.22	1.84	1.89	1.58
Flour, cereal	0.43	0.43	.11	1.02	0.21
Meat	5.25	5.16	4.62	5.99	5.25
Poultry, fish	7.38	8.93	8.00	7.44	4.39
Eggs	7.51	8.70	6.86	5.92	4.27
Sugar, sweets	3.31	2.97	3.95	3.53	3.02
Potatoes, sweet potatoes	9.30	5.03	12.66	12.88	6.67
Fresh vegetables[a]	21.20	13.51	30.86	18.52	13.66
Fresh fruit[a]	8.11	6.71	11.71	7.19	5.00
Juice (vegetable, fruit)	1.92	2.27	2.52	2.49	2.34
Dried vegetables, fruits	4.44	4.26	12.50	4.17	2.08

[a] Higher total consumption in summer than other seasons.

Source: Dietary consumption distributions of selected food groups for the United States (2).

REFERENCES

1. U.S. Department of Agriculture. Food
intakes: individuals in 48 states, year
1977-78. Washington, DC: USDA, 1980;
Nationwide Food Consumption Survey,
1977-78, Report No. I-1.

2. Purdue Research Foundation. Dietary
consumption distributions of selected
food groups for the U.S. population.
Washington, DC: Environmental
Protection Agency, 1980;
EPA 560/11-80-012 and PB81-147035.

Foodstuffs included in the food groups used in Appendix E table headings are defined here.

Table E.1.
Meat, Fish, and Poultry

Beef

Includes beef steaks, roasts, ground beef, baby-food beef, beef bacon, pastrami, oxtails, and shortribs.

Excludes variety meats, such as liver and kidney, and processed beef, such as beef bologna and beef frankfurters.

Pork

Includes ham, bacon, salt pork, pigs' feet, pork cracklings, baby-food pork, and fresh, cured, smoked, and salted pork.

Excludes variety meats and frankfurters, sausages, and luncheon meats.

Lamb, veal, game

Includes lamb, veal, goat, mutton, baby-food lamb, rabbit, venison, and other game.

Excludes variety meats.

Total poultry

Includes chicken, turkey, duck, goose, cornish game hen, quail, pheasant, other wildfowl, and baby-food chicken.

Excludes giblets.

Chicken

Includes chicken only.

Excludes giblets.

Organ meats, mixtures

Includes liver, heart, kidney, and other organ meats from beef, pork, lamb, veal, game, and poultry; baby-food liver and heart; and mixtures (mainly organ meat).

Frankfurters, sausages

Includes processed meats from beef, and luncheon meats from pork, ham, veal, chicken, and turkey.

Fish, shellfish

Includes finfish; shellfish such as oysters, clams, crabs, lobster, scallops, and shrimp; and other seafood, including frog, fish roe, squid, and turtle.

Mixtures (mainly meat)

Includes mixtures reported as a single poultry, fish unit—stews, casseroles, pot pies, soups, salads, hash, frozen plate meals, meat gravies, and sandwiches when reported as a single item (e.g., ham sandwich).

Table E.2.
Milk, Milk Products, Eggs, Legumes, Nuts, and Seeds

Total Milk, Milk Products

Includes milk, milk drinks, cream, milk desserts, and cheese.

Excludes butter. Milk sauces and gravies are included in this total.

Total milk, milk drinks

Includes fluid milk (see next group for inclusions), yogurt (including frozen), chocolate milk, milk shakes, other milk drinks, liquid meal replacements with milk, and milk-based baby formulas.

Fluid milk

Includes whole, lowfat, skim, acidophilus, soy-based, filled evaporated, and condensed milk; buttermilk; goat milk; and reconstituted dry milk.

Cream, milk desserts

Includes fluid and powdered cream, half-and-half, sour cream, ice cream, ice milk, milk sherbets, and desserts made with milk, such as custards, cornstarch pudding, and baby-food puddings.

Excludes nondairy cream substitutes, which are included under fats and oils.

Cheese

Includes natural hard and soft cheeses, processed cheeses and spreads, imitation cheeses, cottage cheese, cream cheese, and mixtures (mainly cheese), such as cheese souffle, rarebit, and, if reported as a single item, cheese sandwich.

Eggs

Includes whole eggs, egg whites and yolks, baby-food egg yolks, egg substitutes, and mixtures mainly egg, such as omelets, egg salad, and egg sandwiches reported as a single item.

Legumes, nuts, seeds

Includes cooked dry beans, peas, and lentils, mixtures (mainly legumes), such as baked beans and soups; soybean-derived products, such as soy-based baby formulas and imitation milk; frozen meals with cooked dry beans or peas as the main course; meat substitutes (mainly vegetable protein); nuts; peanut butter; seeds; and carob products.

Table E.3.
Grain Products, Fats, and Oils

Bread, rolls, biscuits

Includes all types of yeast breads and rolls, sweet rolls, yeast-type coffee cakes, English muffins, biscuits, and bagels.

Excludes quick breads such as cornbread.

Other baked goods

Includes cornbread, tortillas, plain and fruit muffins, and other quick breads, cakes, cookies, pies, pastries, doughnuts, crackers, salty snacks made from grain products, pancakes, waffles, and French toast.

Cereals, pastas

Includes macaroni, noodles, spaghetti, ready-to-eat and cooked cereals, grits, rice, and other cooked cereals and grains.

Mixtures (mainly grain)

Includes mixtures (some with small amounts of meat and others without meat) such as pizza, enchiladas, spaghetti with sauce, quiche, egg rolls, rice and pasta mixtures, frozen meals with the main course mainly grain, and noodle and rice soups.

Fats, Oils

Includes table fats, cooking fats such as bacon grease, lard, and meat drippings; vegetable oils; salad dressings; nondairy sour cream and sweet cream substitutes; and hollandaise and other sauces mainly fat or oil.

Table fats

Includes butter, margarine, and imitation margarine.

Salad dressings

Includes mayonnaise and regular and low-calorie salad dressings.

Table E.4.
Vegetables

White Potatoes
Includes baked, boiled, mashed, fried, and canned potatoes; potato chips; and mixtures (mainly potato), such as potato salad and potato soup.

Excludes viandas (Puerto Rican starchy vegetables).

Tomatoes
Includes raw and cooked tomatoes; tomato juice and soup; catsup, chili sauce, and other tomato sauces; and mixtures such as tomato and corn, tomato and okra, and tomato sandwich reported as a single item.

Dark-green vegetables
Includes raw and cooked dark-green, leafy vegetables such as chard, collards, escarole, mustard and turnip greens, kale, and spinach; broccoli; mixtures (mainly dark-green vegetables) such as spinach souffle and spinach soup; and baby-food spinach.

Deep-yellow vegetables
Includes raw and cooked deep-yellow or orange vegetables—carrots, pumpkin, winter squash, and sweet potatoes; mixtures (mainly deep-yellow vegetables), such as peas and carrots and sweet potato casserole; and baby-food carrots, squash, and sweet potato.

Other vegetables
Includes cooked and raw vegetables other than white potatoes, tomatoes, dark-green and deep-yellow vegetables and their mixtures, vegetable juices and soups; pickles, olives, and relishes; salad viandas (Puerto Rican starchy vegetables); baby-food vegetable mixtures with meat; and mixtures (mainly vegetables).

Table E.5.
Fruits

Total citrus fruits, juices
Includes oranges and other citrus fruits, orange juice and other citrus juices, mixtures of citrus and other fruit juices, baby-food citrus juices.

Excludes citrus fruit ades and drinks such as lemonade, which are tabulated under fruit drinks and ades.

Dried fruits
Includes apples, apricots, figs, prunes, raisins, and other dried fruits.

Excludes juices such as prune juice and mixtures.

Total other fruits, mixtures
Includes raw and cooked apples, bananas, juice berries, and other fruits except citrus and dried fruit; fruit salads and mixtures (mainly fruit); noncitrus juices including prune juice and nectars; and baby-food noncitrus fruits, juices, and nectars.

Excludes fruit drinks and ades.

Apples
Includes raw and cooked apples, applesauce, and baby-food applesauce.

Bananas
Includes raw and cooked bananas and baby-food bananas.

Other fruits (mainly mixtures)
Includes fruits other than citrus fruits; dried fruits, apples, and bananas; and baby-food noncitrus fruits and mixtures.

Noncitrus juices, nectars
Includes fruit juices other than citrus and baby-food noncitrus juices.

Excludes noncitrus fruits, drinks, and ades.

Table E.6.
Sugar, Sweets, and Beverages

Total sugar, sweets
Includes sugar, sugar substitutes, syrups, honey, molasses, icing, topping, sweet sauces, jelly, jam, marmalade, preserves, sweet pastes, fruit butters, gelatin desserts, ices, popsicles, and candy (including dietetic sweets).

Sugar
Includes white, brown, maple, and raw sugar and sugar substitutes.

Candy
Includes candy (including dietetic sweets), chewing gum, and cough drops.

Total nonalcoholic beverages
Includes coffee, tea, soft drinks, and fruit drinks and ades.

Coffee
Includes ground and instant decaffeinated and regular coffee, coffee mixes, and coffee substitutes.

Tea
Includes tea from leaves; instant tea; and instant tea with lemon, cream, milk, sugar, and/or artificial sweetener; and herbal and other teas.

Soft drinks
Includes carbonated drinks, such as colas, fruit-flavored and cream sodas, ginger ale, root beer, and carbonated diet drinks; and noncarbonated diet drinks; and noncarbonated soft drinks made from powdered mixes and liquid concentrates.

Fruit drinks, ades
Includes regular and low-calorie fruit drinks, punches, and ades, including those made from powdered mix and liquid concentrate.

Total alcoholic beverages
Includes cocktails, and other mixed drinks, liqueurs, wine, distilled liquors, beer, and ale.

Beer, ale
Includes beer, ale, and low-calorie beer.

Table E.7.
Home-Grown Consumption

This table is based on findings of a 1966 USDA food consumption survey of home-grown foods. Approximately 44 percent of U.S. households with gardens have seeded areas of greater than 750 square feet. It has been estimated that 6 million acres are devoted to home gardens. This is approximately equivalent to the acreage devoted to commercial production of fruits and vegetables, so a significant portion of the U.S. diet is derived from home-grown produce.

Milk, Cream, and Cheese
Includes milk, buttermilk, goat milk, milk drinks, cream, sour cream, ice cream, ice milk, milk sherbets, and desserts made with milk, such as custards, cornstarch pudding, and baby-food puddings and cheese (natural hard and soft cheeses, spreads, cottage cheese, cream cheese, and mixtures [mainly cheese], such as cheese souffle, rarebit, and if reported as a single item, cheese sandwich).

Excludes butter.

Fats, Oils
Includes table fats, cooking fats such as bacon grease, lard, and meat drippings; vegetable oils; salad dressings; and hollandaise and other sauces mainly fat or oil.

Flour and cereal
Includes all types of yeast breads and rolls, sweet rolls, yeast-type coffee cakes, muffins, biscuits, bagels, cornbread, tortillas, plain and fruit muffins, other quick breads, cakes, cookies, pies, pastries, doughnuts, crackers, salty snacks made from grain products, pancakes, waffles, french toast, macaroni, noodles, spaghetti, cooked cereals, grits, rice, and other cooked cereal grains.

Meat

Includes beef steaks, roasts, ground beef, baby-food beef, beef bacon, pastrami, oxtails, shortribs, ham, bacon, salt pork, pigs' feet, pork cracklings, fresh pork, cured pork, smoked pork, salted pork, lamb, veal, goat, mutton, rabbit, venison, and other game.

Poultry and fish

Includes chicken; turkey; duck; goose; cornish game hen; quail; pheasant; other wildfowl; finfish; shellfish such as oysters, clams, crabs, lobster, scallops, and shrimp; and other seafood, including frog, fish roe, squid, and turtle.

Eggs

Includes whole eggs, egg whites and yolks, mixtures (mainly egg), such as omelets, egg salad, and egg sandwiches reported as a single item.

Sugar and sweets

Includes sugar, sugar substitutes, syrups, honey, molasses, icing, topping, sweet sauces, jelly, jam, marmalade, preserves, sweet pastes, fruit butters, gelatin desserts, ices, popsicles, and candy.

Potatoes and sweet potatoes

Includes baked, boiled, mashed, fried, and canned potatoes; potato chips; and mixtures (mainly potato), such as potato salad and potato soup; sweet potatoes, and sweet potato casserole.

Excludes viandas (Puerto Rican starchy vegetables).

Fresh vegetables

Includes raw and cooked tomatoes; tomato juice and soup; catsup, chili sauce, and other tomato sauces; and mixtures such as tomato and corn, tomato and okra, and tomato sandwich reported as a single item; raw and cooked dark-green leafy vegetables such as chard, collards, escarole, mustard and turnip greens, kale, and spinach; broccoli; mixtures (mainly dark-green vegetables), such as spinach souffle and spinach soup; baby-food spinach; raw and cooked deep-yellow or orange vegetables such as carrots, pumpkin, winter squash, and sweet potatoes; mixtures (mainly deep-yellow vegetables), such as peas and carrots and sweet potato casserole; baby-food carrots, squash, and sweet potatoes; cooked and raw vegetables other than white potatoes, tomatoes, dark-green and deep-yellow vegetables and their mixtures, vegetable juices and soups; pickles, olives, and relishes; salad viandas (Puerto Rican starchy vegetables).

Fresh fruit

Includes oranges and other citrus fruits, raw and cooked apples, bananas, juices, berries, fruit salads and mixtures (mainly fruit), applesauce, and other fruits.

Juice (vegetable and fruit)

Includes vegetable juices such as tomato; citrus and noncitrus juices, drinks, ades, and nectars.

Dried vegetables and fruits

Includes onions, apples, apricots, figs, prunes, raisins, and other dried fruits.

Health outcome data sources may be available in various formats, including computerized tapes or disks and hard copy. The majority of these sources can be identified through state health departments when their use is appropriate; they are all options.

I. STATE

A. Vital Statistics: State health department

- Birth Certificates

- Death Certificates

- Fetal Death Reports

B. Registries: State health department

1. Disease

    • Cancer/Tumor

    • Congenital Malformation/Birth Defects

2. Exposure

    • Occupational Exposure

    • Environmental Exposure

C. Health Studies

The state health department usually knows of health studies that have involved the site population. These studies may have been conducted at the federal, state, (including universities), and local levels and may include the following types:

- Symptom or Disease Prevalence Surveys

- Exposure Studies

- Cluster Investigations

- Analytic (Epidemiologic) Studies

D. Complaint Records

The environmental sections of some state health departments keep logs of complaints from citizens. For example, Michigan and Connecticut maintain logs of complaints regarding air pollution. Michigan also maintains a record of complaints regarding toxic substances.

E. School Attendance Records

These are located at the state education department. The State Department of Education can direct you to local sources.

F. Census Data

- State census data center of Government Planning Office

II. LOCAL SOURCES

A. Health Records: local health department

- Complaint logs

- Health Studies (see number I.C.)

B. Medical Records

Although medical records and other information with personal identifiers may be sources of health outcome data, this type of information should not be routinely collected when conducting health assessments. Examples of medical records include:

- Hospital Discharge Records

- Hospital Emergency Room Logs

- Private Physicians' Records

- School Nurses' Records

- Health Clinics: Occupational or Free-Standing

- Facility Occupational Health Records

Note: None of the following data sources are site-specific, but they may be used for comparison purposes.

I. DEPARTMENT OF HEALTH AND HUMAN SERVICES

A. National Center for Health Statistics

1. National Ambulatory Care Survey - Survey of information on provision and use of ambulatory medical care services. Data includes the primary reason for visit, physician's primary diagnosis, treatment, prior visits, and final disposition. (Years available: 1973 - 1981).

2. Vital Statistics - Birth and death records in a national registry.

3. National Health Interview Survey - United States survey conducted through household interviews. Data include the number of restricted activity days, the number of bed days, lost work/school days, and the number of visits to physicians or dentists.

4. National Health and Nutrition Examination Survey (NHANES) - United States cross sample of individuals. The data collected include dietary, hematological, and biochemical (limited information on pesticides, metals, and VOC levels in the general population), body measurements, and clinical assessment.

B. National Cancer Institute (NCI)

1. Surveillance, Epidemiology, and End Result (SEER) - The American Cancer Society issues an annual report of cancer statistics using the incidence data generated by the SEER program.

2. Riggan's Mortality Tapes - Produced by NCI and EPA. These tapes provide a comparison of the number of deaths from a specific cancer type in a specified county and state with the number of deaths from the same cancer for the entire United States.

C. NIOSH

- Information on occupationally related exposures, illnesses, and accidents.

D. ATSDR

- Exposure Registry and Disease Registries

- Health Studies

E. Other DHHS Agencies (e.g., NIEHS)

II. U.S. BUREAU OF CENSUS

National population and housing data

HEALTH OUTCOME DATA CHECKLIST

The following Checklist was developed to assist the Health Assessor in characterizing Health Outcome Data Sources at the federal, state, and local levels. A checklist needs to be completed for <u>each</u> data source identified for a site. Many of the variable will have yes or no (y/n) responses. The Health Assessor should note any deviations from the checklist. Definitions of terms used in the checklist are provided in the attached glossary.

Preparer's Name: _____

Date Prepared: _____

Site Name: _____

Site Address:
           Street _____
           City/Town _____
           County _____
           State _____

Primary Contact Person: _____
Agency/Affiliation: _____
Phone Number: _____

TYPE: VITAL STATISTICS (Check One)

___ Birth Certificate
___ Death Certificate
___ Fetal Death Report

** Obtain a copy of the Blank Certificate or Report Form **

Agency Responsible for Reporting the data:	Agency Responsible for Maintaining the data?
___ Federal Name: _____ Years data reported? ____ to ____ QA/QC documented for the data source? Yes ____ or No ____	___ Federal Name: _____ How frequently are the data updated? ____ Monthly ____ Quarterly ____ Yearly
___ State Name: _____ Years data reported? ____ to ____ QA/QC documented for the data source? Yes ____ or No ____	___ State Name: _____ How frequently are the data updated? ____ Monthly ____ Quarterly ____ Yearly
___ Local Name: _____ Years data reported? ____ to ____ QA/QC documented for the data source? Yes ____ or No ____	___ Local Name: _____ How frequently are the data updated? ____ Monthly ____ Quarterly ____ Yearly

The following types of data forms are available and for what years?	For what geographic areas are the data available and for what years?	Please check all Data Variables and years included in the data source.
Report Form: Years Available: ___ Computerized ____ ___ Tape/cartridge ____ ___ Diskette ____ ___ Other ____ Specify ____	Areas: Years Available: ___ Region ____ ___ State ____ ___ County ____ ___ City/Town/Twnshp ____ ___ Zip Code ____ ___ Census Tract ____ ___ Block, ED, BG ____	Variables available? Years: ___ Age ____ ___ Sex ____ ___ Ethnicity ____ ___ Address ____
___ Hard Copy ____ ___ Summary Report ____ ___ Other ____ Specify ____	Are there any Confidentiality Issues for the data? Yes ____ or No ____ Specify _____	

COMPLETE THE APPROPRIATE BOX FOR THE DATA SOURCE:

Birth Certificate:	Death Certificate:	Fetal Death Report:
___ Birthweight ___ Birth Defects ___ Gestational Age ___ Other ____	___ Immediate Cause of Death ___ Underlying Cause of Death ___ Other Significant Condition	___ Immediate Cause of Death ___ Underlying Cause of Death ___ Other Significant Condition ___ Fetal Weight ___ Gestational Age ___ Congenital Malformations

Contact Person: _____
Phone Number: _____

TYPE: REGISTRIES (Check One)

____ Disease (Specify)   ____ Exposure
____ Cancer   Specify: _____
____ Tumor
____ Congenital Malformation, Birth Defects
____ Other: _____

Agency Responsible for Reporting the data:	Agency Responsible for Maintaining the data?
___ Federal Name: _____ Years data reported? _____ to _____ QA/QC documented for the data source? Yes _____ or No _____	___ Federal Name: _____ How frequently are the data updated? _____ Monthly _____ Quarterly _____ Yearly
___ State Name: _____ Years data reported? _____ to _____ QA/QC documented for the data source? Yes _____ or No _____	___ State Name: _____ How frequently are the data updated? _____ Monthly _____ Quarterly _____ Yearly
___ Local Name: _____ Years data reported? _____ to _____ QA/QC documented for the data source? Yes _____ or No _____	___ Local Name: _____ How frequently are the data updated? _____ Monthly _____ Quarterly _____ Yearly

For what geographic areas are the data reported, and for what years?	For what geographic areas can I get rates (morbidity and mortality) and for what years?

Areas: Years Reported:
 	
____ Entire State
____ Portion of State _____
 	
How much (percentage)? _____
 	
Which Geographic Areas:
_____ _____
_____ _____
_____ _____
	Areas: Years Available:
	
	___ Region
	    Specify: _____ _____
	___ State _____
	___ County _____
	    Specify: _____
	___ City/Town/Twnshp _____
	    Specify: _____
	___ Zip Code _____
	___ Census Tract _____
	___ Block, ED, BG _____

Registries: Page 2

The following types of data forms are available and for what years

Report Form: Years Available: Report Form: Years included in data source:
                                               Hard Copy   _____
___ Computerized   _____   ___ Summary Report   _____
___ Tape/cartridge   _____   ___ Other
___ Diskette   _____
___ Other   _____   Specify _____

    Specify _____

                                               Please check all the measures of effect reported in
                                               the hard-copy data, and for what years.

Please check all the variables available in the   Measure of Effect: Years for which reported:
computerized data source, and for what years.   ___ Incidence   _____
                                                     ___ Crude   _____
    Data Variables: Years included: ___ Age-adjusted   _____
    ___ Age   _____   ___ Sex-Adjusted   _____
    ___ Sex   _____   ___ Race-Adjusted   _____
    ___ Ethnicity   _____   ___ Odds Ratio   _____
    ___ Address   _____   ___ Crude   _____
    ___ Diagnosis   _____   ___ Age-Adjusted   _____
                                                     ___ Sex-Adjusted   _____
                                                     ___ Race-Adjusted   _____

Who is the person (or organization) who reports the   What is the completeness of reporting by this person
data to the data source?   (organization)?

___ Hospitals
    Specify: _____   ___ Complete
___ Private Physicians
    Specify: _____   ___ Partial
___ Laboratories
    Specify: Specify Percentage: _____

Are there any Confidentiality Issues for the data? Yes _____ or No _____

    Specify _____
            _____

Contact Person: _____

Address/Affilaiation: _____

Phone: _____

TYPE: STUDIES (Check One)

____ Symptom/Disease Prevalence
____ Exposure
____ Cluster Investigation
____ Analytic Studies

**** Please complete a separate form for each study, and obtain a copy of study proposal and findings.

What is the geographic area included in the study? Specify: _____ _____ _____	Who is the Agency Responsible for Conducting the study (e.g. data collection)?
For what years were the data collected? Specify: _____ to _____	____ Federal Name: _____ Who is the primary investigator for the study? Name: _____ Address/Affiliation: _____ Phone: _____
	____ State (include universities) Name: _____ Who is the primary investigator for the study? Name: _____ Address/Affiliation: _____ Phone: _____
	____ Local Name: _____ Who is the primary investigator for the study? Name: _____ Address/Affiliation: _____ Phone: _____

Contact Person: _____

Address/Affiliation: _____
_____

Phone: _____

TYPE: RECORDS (Checklist One)

_____ Medical (e.g. Hospital, Physician, ER Log)
_____ Educational (Attendance)
_____ Compliant (e.g. Air Pollution)
_____ Occupational Health

Agency Responsible for Reporting the data:	Agency Responsible for Maintaining the data?
__ Federal Name: _____ Years data reported? _____ to _____ QA/QC documented for the data source? Yes _____ or No _____	__ Federal Name: _____ How frequently are the data updated? _____ Monthly _____ Quarterly _____ Yearly
__ State Name: _____ Years data reported? _____ to _____ QA/QC documented for the data source? Yes _____ or No _____	__ State Name: _____ How frequently are the data updated? _____ Monthly _____ Quarterly _____ Yearly
__ Local Name: _____ Years data reported? _____ to _____ QA/QC documented for the data source? Yes _____ or No _____	__ Local Name: _____ How frequently are the data updated? _____ Monthly _____ Quarterly _____ Yearly

The following types of data forms are available and for what years?	For what geographic areas are the data available and for what years?	Who is the person (or organization) who reports the data to the data source?
Report Form: Years Available: ___ Computerized _____ ___ Tape/cartridge _____ ___ Diskette _____ ___ Other _____ Specify _____	Areas: Years Available: ___ Region _____ ___ State _____ ___ County _____ ___ City/Town/Twnshp _____ ___ Zip Code _____ ___ Census Tract _____ ___ Block, ED, BG _____	_____ Hospitals Specify: _____ _____ Private Physicians Specify: _____ _____ Laboratory Specify: _____

___ Hard Copy _____
___ Summary Report _____
___ Other _____
 	
Specify _____
	Are there any Confidentiality Issues for the data?
	Yes _____ or No _____
	  Specify: _____
	  _____
	  _____
		What is the completeness of reporting by this person or organization?
		_____ Complete
		
		_____ Partial
		
		Specify the percentage: _____

RECORDS: Page 2

Please check all Data Variables included in the data source.	For what years are these variables available?
_____ Age _____ Sex _____ Ethnicity _____ Address	_____ _____ _____ _____

COMPLETE THE APPROPRIATE BOX:

Medical: Record Type: ___ Hospital Discharge ___ Physician's Records ___ Hospital Emergency Room Logs ___ Other: _____	Educational: Geographic Area covered by the data: _____ State _____ District Name: _____ _____ School Name: _____	Complaints: ___ Type Specify: _____ Frequency: _____ Daily _____ Weekly _____ Monthly
Name of the facility responsible for preparing the Medical Record? _____ Address: _____	_____ Total Enrollment _____ Absentee Rates _____ Other Specify: _____ _____	What is the Geographic Area covered by the data? _____ _____
Please check which of the following data are available. _____ Diagnosis _____ Underlying Conditions _____ Risk Factors		Do the data indicate the type of follow-up provided for the complaints? Yes _____ or No _____

Contact Person: _____

Address: _____

Phone: _____

SOURCES OF
HEALTH
OUTCOME
DATA

SOURCES OF DATA FOR ANALYSIS OF HEALTH OUTCOMES POTENTIALLY AVAILABLE FOR SITE- SPECIFIC POPULATIONS

Routinely collected health data	Vital statistics records Registries Hospital/medical records School records
Routinely collected population data	Census data
Non-routine health data	Previously conducted health studies

Vital Statistics Records

Death Certificate

Natality Records:
   Birth Certificate
   Fetal Death Certificate

- Completion is a legal requirement

- Reporting administered by National Center for Health Statistics (NCHS)

- Complete and high quality reporting across U.S. since 1930's

VITAL STATISTICS REGISTRATION SYSTEMS IN THE U.S.

Responsible Person or Agency	Death Certificate	Birth Certificate	Fetal Death Certificate
Physician, other attendant	Completes/ signs medical certification. Sends to funeral director	Completes/ signs medical certificate. Filed with local agency	Completes/ signs medical certification. Sends to funeral director or local agency
Funeral Director	Personal facts. Delivers completed certificate to local agency	xxx	Obtains personal facts Delivers completed certificate to local agency
Local agency (Registrar or Hlth Dept)	Verifies completeness of certificates Maintains records for local use and reports Sends certificate to state agency		
State agency, Bureau of Vital Statistics	Queries incomplete or inconsistent information Maintains records for state use and reports Transmits records to NCHS		
NCHS	Maintains national records Publishes national statistical reports Maintains technical assistance for quality assurance		

DEATH CERTIFICATE INFORMATION SYSTEMS MAINTAINED BY VITAL STATISTICS BUREAUS

Note: Some information may be confidential and only accessible in summary form

Local, State, Federal
Hard copy, microfilm/fiche of certificates

State, Federal and Some Local
Computer data tapes
- By calendar year of death
- Major items coded and entered for each death

Causes of Death Coded Using the International Classification of Disease (ICD)
- Underlying cause of death only (< 1968)
- Multiple causes of death (1968-   )

State, Federal and Some Local
Published vital statistics reports

DEATH CERTIFICATE

insert attached

DEPARTMENT OF HEALTH, EDUCATION, AND WELFARE–PUBLIC HEALTH SERVICE–NATIONAL CENTER FOR HEALTH STATISTICS
1978 REVISION

(PHYSICIAN, MEDICAL EXAMINER OR CORONER)
U.S. STANDARD
CERTIFICATE OF DEATH

Form Approved
OMB No. 68R 1901

LOCAL FILE NUMBER — STATE FILE NUMBER

TYPE OR PRINT IN PERMANENT INK
FOR INSTRUCTIONS SEE HANDBOOK

DECEDENT

DECEDENT–NAME FIRST MIDDLE LAST 1.			SEX 2.	DATE OF DEATH (Mo., Day, Yr.) 3.
RACE–(e.g., White, Black, American Indian, etc.) (Specify) 4.	AGE–Last Birthday (Yrs.) 5a.	UNDER 1 YEAR: MOS. DAYS 5b. UNDER 1 DAY: HOURS MINS. 5c.	DATE OF BIRTH (Mo., Day, Yr.) 6.	COUNTY OF DEATH 7a.
CITY, TOWN OR LOCATION OF DEATH 7b.	HOSPITAL OR OTHER INSTITUTION–Name (If not in either, give street and number) 7c.			IF HOSP. OR INST. Indicate DOA, OP/Emer. Rm., Inpatient (Specify) 7d.
STATE OF BIRTH (If not in U.S.A., name country) 8.	CITIZEN OF WHAT COUNTRY 9.	MARRIED, NEVER MARRIED, WIDOWED, DIVORCED (Specify) 10.	SURVIVING SPOUSE (If wife, give maiden name) 11.	WAS DECEDENT EVER IN U.S. ARMED FORCES? (Specify Yes or No) 12.
SOCIAL SECURITY NUMBER 13.	USUAL OCCUPATION (Give kind of work done during most of working life, even if retired) 14a.		KIND OF BUSINESS OR INDUSTRY 14b.	

IF DEATH OCCURRED IN INSTITUTION, SEE HANDBOOK REGARDING COMPLETION OF RESIDENCE ITEMS

RESIDENCE–STATE 15a.	COUNTY 15b.	CITY, TOWN OR LOCATION 15c.	STREET AND NUMBER 15d.	INSIDE CITY LIMITS (Specify Yes or No) 15e.

PARENTS

FATHER–NAME FIRST MIDDLE LAST 16.	MOTHER–MAIDEN NAME FIRST MIDDLE LAST 17.
INFORMANT–NAME (Type or Print) 18a.	MAILING ADDRESS STREET OR R.F.D. NO. CITY OR TOWN STATE ZIP 18b.

DISPOSITION

BURIAL, CREMATION, REMOVAL, OTHER (Specify) 19a.	CEMETERY OR CREMATORY–NAME 19b.	LOCATION CITY OR TOWN STATE 19c.
FUNERAL SERVICE LICENSEE Or Person Acting As Such (Signature) 20a. ▶	NAME OF FACILITY 20b.	ADDRESS OF FACILITY 20c.

CERTIFIER

To be Completed by CERTIFYING PHYSICIAN Only

21a. To the best of my knowledge, death occurred at the time, date and place and due to the cause(s) stated.
(Signature and Title) ▶

DATE SIGNED (Mo., Day, Yr.) 21b.	HOUR OF DEATH 21c. M
NAME OF ATTENDING PHYSICIAN IF OTHER THAN CERTIFIER (Type or Print) 21d.	

To be Completed by MEDICAL EXAMINER or CORONER Only

22a. On the basis of examination and/or investigation, in my opinion death occurred at the time, date and place and due to the cause(s) stated.
(Signature and Title) ▶

DATE SIGNED (Mo., Day, Yr.) 22b.	HOUR OF DEATH 22c. M
PRONOUNCED DEAD (Mo., Day, Yr.) 22d. ON	PRONOUNCED DEAD (Hour) 22e. AT M

NAME AND ADDRESS OF CERTIFIER (PHYSICIAN, MEDICAL EXAMINER OR CORONER) (Type or Print)
23.

REGISTRAR 24a. (Signature) ▶	DATE RECEIVED BY REGISTRAR (Mo., Day, Yr.) 24b.

CAUSE OF DEATH

CONDITIONS IF ANY WHICH GAVE RISE TO IMMEDIATE CAUSE STATING THE UNDERLYING CAUSE LAST

25. IMMEDIATE CAUSE (ENTER ONLY ONE CAUSE PER LINE FOR (a), (b), AND (c).)

PART I
(a) — Interval between onset and death
DUE TO, OR AS A CONSEQUENCE OF:
(b) — Interval between onset and death
DUE TO, OR AS A CONSEQUENCE OF:
(c) — Interval between onset and death

PART II OTHER SIGNIFICANT CONDITIONS–Conditions contributing to death but not related to cause given in PART I (a)		AUTOPSY (Specify Yes or No) 26.	WAS CASE REFERRED TO MEDICAL EXAMINER OR CORONER (Specify Yes or No) 27.
ACC., SUICIDE, HOM., UNDET., OR PENDING INVEST. (Specify) 28a.	DATE OF INJURY (Mo., Day, Yr.) 28b.	HOUR OF INJURY 28c. M	DESCRIBE HOW INJURY OCCURRED 28d.
INJURY AT WORK (Specify Yes or No) 28e.	PLACE OF INJURY–At home, farm, street, factory, office building, etc. (Specify) 28f.	LOCATION STREET OR R.F.D. No. CITY OR TOWN STATE 28g.	

HRA-162-1
Rev. 1/78

INFORMATION OF RELEVANCE ON DEATH CERTIFICATES

A standard certificate is prepared by NCHS, but some items vary by state.

Health information:
Completed by physician or legal authority such as coroner, medical examiner

Part I Immediate Cause: Enter only one cause per line

(a) _____

Due to or as a consequence of

(b) _____

Due to or as a consequence of

(c) _____

Part II Other significant conditions	Autopsy Yes/No

Acc, suicide homicide or pending	Date of injury	Hour of injury	How injury occurred

Assigning Cause of Death:

Line:

(a) Immediate cause

(b) Antecedent condition, if any, that gave rise to (a)

(c) Antecedent condition, if any, that gave rise to (a) and (b)

The underlying cause, the last listed condition, is the disease or injury that initiated the sequence of morbid events leading directly or indirectly to death.

OTHER RELEVANT INFORMATION ON DEATH CERTIFICATES

Date of death

Age at death

Sex

Race

Residence - state, county, city, street address

Marital status

Usual occupation, kind of business/industry

Place of death - town, hospital

INTERNATIONAL CLASSIFICATION OF DISEASES (ICD)

Standardized coding system for causes of death and morbidity under auspices of World Health Organization.

ICD Code	Major Category
001-139	Infectious disease
140-239	Neoplastic diseases
240-279	Endocrine, metabolic immune disorders
280-289	Diseases of blood and blood organs
290-319	Mental disorders
320-389	Diseases of central nervous system
390-459	Diseases of circulatory system
740-759	Congenital anomalies
760-779	Conditions originating in perinatal period
780-799	Ill-defined conditions
800-999	Accidents, injuries and poisoning

MAJOR USE OF DEATH CERTIFICATE DATA

• Analysis of patterns of causes of death for residents of geographic areas

Advantages of Analysis of Death Certificate Data:
• Economical and efficient
• All deaths registered
• Comparisons among local areas, states and nationally
• Comparisons by sex, age, race, time period
• Available over many decades
• Good representation of patterns for diseases that are highly, rapidly fatal and readily diagnosed

Disadvantages of Analysis of Death Certificate Data:
• Inaccuracy of physician's assignment of cause of death and ICD coding
• Local and temporal variations in physicians' practice of assigning cause of death
• Poor representation of patterns for diseases that are not highly and rapidly fatal or readily diagnosed
• Inadequate for conditions which do not cause death
• Disaggregation of deaths to small geographic areas may not be possible from computerized data
• Residence based on last residence (at time of death)

BIRTH CERTIFICATE INFORMATION SYSTEMS MAINTAINED BY VITAL STATISTICS BUREAUS

Note: Some information may be confidential and only accessible in summary form

Local, State, Federal

Hard copy, microfilm/fiche of certificates

State, Federal and some Local

Computer data tapes
- By calendar year of deaths
- Major items coded and entered for each birth
- Congenital abnormalities coded by ICD-CM

State, Federal and Some Local

Published vital statistics reports

INFORMATION OF RELEVANCE ON BIRTH CERTIFICATE

A standard certificate is prepared by NCHS, but some items vary by state authority.

Health information:

Completed by physician or non-physician attendant

Birth weight - (low birthweight)

Estimated gestational age - (premature)

APGAR scores

Congenital malformations or anomalies

Complications of pregnancy

Mother's reproductive history

Demographic Information:

Date of Birth

Sex

Race/Hispanic origin

Residence of mother at time of birth - state, county, town, street address

Age of mother, father

Occupation of mother, father

CERTIFICATE OF LIVE BIRTH

Insert attached

TYPE
OR PRINT
IN
PERMANENT
INK
FOR
INSTRUCTIONS
SEE
HANDBOOK

CHILD

CERTIFIER

MOTHER

FATHER

Form Approved
OMB No. 68R 1900

U.S. STANDARD
CERTIFICATE OF LIVE BIRTH

LOCAL FILE NUMBER

BIRTH NUMBER

CHILD—NAME	FIRST	MIDDLE	LAST	SEX	DATE OF BIRTH (Mo., Day, Yr.)	HOUR
1.				2.	3a.	3b. M

HOSPITAL—NAME (If not in hospital, give street and number)	CITY, TOWN OR LOCATION OF BIRTH	COUNTY OF BIRTH
4a.	4b.	4c.

I certify that the stated information concerning this child is true to the best of my knowledge and belief.	DATE SIGNED (Mo., Day, Yr.)	NAME AND TITLE OF ATTENDANT AT BIRTH IF OTHER THAN CERTIFIER (Type or print)
5a. (Signature) ▶	5b.	5c.

CERTIFIER—NAME AND TITLE (Type or print)	MAILING ADDRESS (Street or R.F.D No., City or Town, State, Zip)
5d.	5e.

REGISTRAR	DATE RECEIVED BY REGISTRAR (Month, Day, Year)
6a. (Signature) ▶	6b.

MOTHER—MAIDEN NAME	FIRST	MIDDLE	LAST	AGE (At time of this birth)	STATE OF BIRTH (If not in U.S.A., name country)
7a.				7b.	7c.

RESIDENCE—STATE	COUNTY	CITY, TOWN OR LOCATION	STREET AND NUMBER OF RESIDENCE	INSIDE CITY LIMITS (Specify yes or no)
8a.	8b.	8c.	8d.	8e.

MOTHER'S MAILING ADDRESS—If same as above, enter Zip Code only
9.

FATHER—NAME	FIRST	MIDDLE	LAST	AGE (At time of this birth)	STATE OF BIRTH (If not in U.S.A., name country)
10a.				10b.	10c.

I certify that the personal information provided on this certificate is correct to the best of my knowledge and belief. (Signature of Parent or other Informant)	RELATION TO CHILD
11a. ▶	11b.

INFORMATION FOR MEDICAL AND HEALTH USE ONLY

RACE—MOTHER (e.g., White, Black, American Indian, etc.) (Specify)	RACE—FATHER (e.g., White, Black, American Indian, etc.) (Specify)	BIRTH WEIGHT	THIS BIRTH—Single, twin, triplet, etc. (Specify)	IF NOT SINGLE BIRTH—Born first, second, third, etc. (Specify)	IS MOTHER MARRIED? (Specify yes or no)
12	13	14	15a.	15b.	16

PREGNANCY HISTORY (Complete each section)				EDUCATION—MOTHER (Specify only highest grade completed)		EDUCATION—FATHER (Specify only highest grade completed)	
				Elementary or Secondary (0-12)	College (1-4 or 5+)	Elementary or Secondary (0-12)	College (1-4 or 5+)
				18		19	

DEATH UNDER ONE YEAR OF AGE Enter State File Number of death certificate for this child

LIVE BIRTHS (Do not include this Child)		OTHER TERMINATIONS (Spontaneous and Induced)		DATE LAST NORMAL MENSES BEGAN (Month, Day, Year)	MONTH OF PREGNANCY PRENATAL CARE BEGAN First, second, etc. (Specify)	PRENATAL VISITS Total number (If none, so state)	APGAR SCORE	
17a. Now living	17b. Now dead	17d. Before 20 weeks	17e. After 20 weeks				1 min	5 min
Number _____	Number _____	Number _____	Number _____	20	21a.	21b.	22a.	22b.

MULTIPLE BIRTHS Enter State File Number for mate(s)

None ☐	None ☐	None ☐	None ☐	COMPLICATIONS OF PREGNANCY (Describe or write "none")
				23

LIVE BIRTH(S)

DATE OF LAST LIVE BIRTH (Month, Year)	DATE OF LAST OTHER TERMINATION (as indicated in d or e above) (Month, Year)	CONCURRENT ILLNESSES OR CONDITIONS AFFECTING THE PREGNANCY (Describe or write "none")
17c.	17f.	24

FETAL DEATH(S)

COMPLICATIONS OF LABOR AND/OR DELIVERY (Describe or write "none")	CONGENITAL MALFORMATIONS OR ANOMALIES OF CHILD (Describe or write "none")
25	26

DEPARTMENT OF HEALTH, EDUCATION, AND WELFARE—PUBLIC HEALTH SERVICE—NATIONAL CENTER FOR HEALTH STATISTICS 1978 REVISION

MAJOR USE OF
BIRTH CERTIFICATE DATA

- Analysis of patterns of health information on newborns to residents of geographical areas

Advantages of Analysis of Birth Certificate Data:

- Economical and efficient
- Birth registration is mostly complete
- Comparisons among local areas, states and nationally
- Comparisons by maternal age, race, time period
- Available over many decades
- Good representation of patterns for
    - Birth weight
    - Severe birth defects readily diagnosed at birth, e.g., anencephaly or spina bifida

Disadvantages of Analysis of Birth Certificate Data:

- Inaccuracies and incompleteness in information such as gestational age, APGAR score, mother's reproductive hx, complications of pregnancy
- Local and temporal variations in physicians' practice of recording of other congenital malformations
- Disaggregation of births to small geographic areas may not be possible from computerized data
- Residence at birth may not be residence throughout pregnancy

FETAL DEATH CERTIFICATES

Fetal Death:

Death prior to complete expulsion or extraction of fetus

Legal certification requirements vary by state:

Most require certification after 20 weeks gestation.

Some require certification regardless of gestation age

Note:

- Certification nearly complete for > 28 weeks gestation
- Certification inconsistent for 20-28 weeks
- Certification incomplete for < 20 weeks

FETAL DEATH CERTIFICATE

A standard certificate is prepared by NCHS, but some items vary by state adoption. Completed by physician or non-physician attendant.

Health Information:

Cause of death - fetal/maternal conditions
Gestational age
Congenital malformations or anomalies

Demographic Information:

As above

USES OF FETAL DEATH CERTIFICATES RESTRICTED BY COMPLETENESS OF REPORTING

(with reporting as noted above)

• Analysis of patterns of fetal deaths > 28 weeks gestation by geographic areas

Disadvantages:

• Incomplete reporting < 28 weeks gestation
• Incomplete reporting of congenital malformations

DISEASE REGISTRIES

- Centralized information collection systems, typically established under the authority of local, state or federal health agencies
- Have a mechanism to identify persons diagnosed with given diseases in their population coverage area

Mechanism:

- Medical care providers may report to registry
- Registry staff may actively review medical records
- Disease diagnosis, demographics and other information collected on each case

Data stored centrally by registry:

- Hard copy, microfilm/fiche of abstracts (forms)
- Computerized data files
- Published reports

CANCER REGISTRIES
(OR TUMOR REGISTRIES)

- Operating in 43 states, District of Columbia and Puerto Rico, where population coverage may be statewide or regional
- Objective is complete identification within their coverage area of all new diagnoses of cancer (typically exclusive of non-melanoma skin cancers) soon after diagnosis (incident cases of cancer)
- Operationally to identify newly diagnosed cancer cases
- May be legal requirement for reporting
- Hospitals may report cases or registry staff may review hospital records
- Typically a strong emphasis placed on completeness of identification of cases and data quality
    - Eliminate duplicate reports
    - Standardized reporting forms
    - Diagnostic criteria
    - Additional checks via death certificates
    - Eliminate non-residents

INFORMATION OF
RELEVANCE ROUTINELY
COLLECTED BY CANCER REGISTRIES

Health Information:

Information is taken from hospital and clinic records including laboratory (pathology, CT scans, x-rays, cytologic) reports. The diagnosis of site/ type of cancer is clinical judgment.

- Site/Type of Cancer
- Staging of Cancer
- ICD-CM code
- Primary vs. Metastatic

Demographic/Other Information:

Date of initial diagnosis

Sex

Age at Dx

Race/Ethnic grouping

Usual Occupation

Residence at Dx: State, County, Town, Street Address

Cancer Surveillance, Epidemiology and End Results Program (SEER)

Operated under auspices of National Cancer Institute and Centers for Disease Control since 1973. SEER aggregates cancer data from eight cooperating registries - (SEER Sites):

California

Connecticut

Georgia

Hawaii

Iowa

Michigan

New Mexico

Utah

Washington

These aggregated data are used for estimate of national cancer incidence.

STATES WITH CANCER REGISTRIES
(State-wide or Regional Coverage)

SEER
Sites
(1973-)

California
Connecticut
Georgia
Hawaii
Iowa
Michigan
New Mexico
Utah
Washington

Registries in 43 states
and in District of
Columbia and Puerto Rico

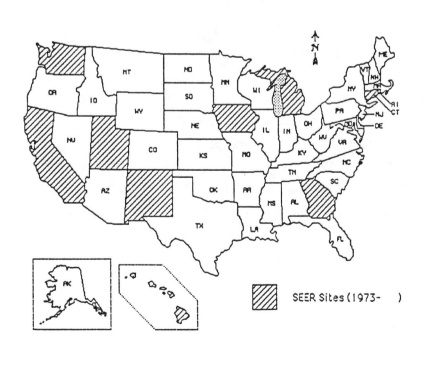

SEER Sites (1973-   )

MAJOR USE OF CANCER REGISTRY DATA:

• Analysis of patterns of newly diagnosed cancer (by type or all) among geographic areas

Advantages of Cancer Registry Data:

• Economical and efficient
• Registry typically complete for severe cancers
• Comparisons by age, sex, race
• Focus is newly diagnosed disease (not just death due to cancer) so not influenced by survival

Disadvantages of Cancer Registry Data:

• Not available in all areas of U.S.
• Time period covered may be short in some areas
• Disaggregation to small localities may not be possible from computerized files
• Accuracy of clinical diagnosis may be unreliable and vary across areas
• Completeness of ascertainment (may vary across areas and over time due to screening for early diagnosis)

Birth Defects Registries

Atlanta, GA: Metropolitan Atlanta Congenital Defects Programs (MACDP) under auspices of CDC (1967-)

Intensive monitoring of the diagnosis of congenital malformations among liveborn or stillborn infants with structural, chromosomal or biochemical abnormality presumed present at birth and diagnosed prior to one year of age.

The operating systems are similar to those of cancer registries, however, the identification of children with birth defects may be incomplete and/or inaccurate as to diagnosis. Some registries may not have long been in operation.

STATES WITH BIRTH DEFECTS REGISTRIES

(State-wide or Regional Coverage)

Active

- Arkansas
- Arizona
- California
- Georgia
- Hawaii
- Iowa
- Washington

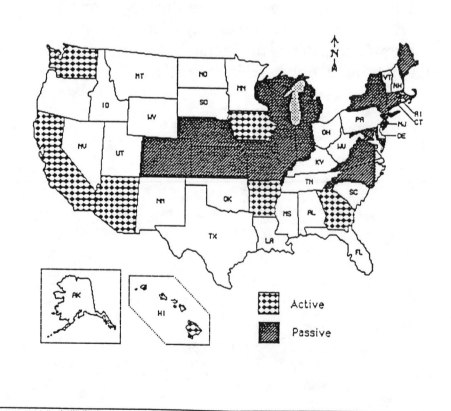

OCCUPATIONAL DISEASE REGISTRIES

Some 60% of states have either mandatory or voluntary reporting programs (reporting to State Health Agency) of selected occupational health conditions, e.g., lead poisoning, silicosis, and asbestosis.

Disadvantages:
- not all industries covered
- reporting may be grossly incomplete and vary in completeness by area
- diagnostic criteria are not standardized

National Institute of Occupational Health and Safety (NIOSH)
- putting into operation a mandatory reporting system nationwide for 10 leading work-related diseases and injuries
- quality of reporting and work-force coverage are yet to be established

MEDICAL RECORDS

Hospital records
- In-patient
- Emergency room

Physician and Clinic Records
School Nurses' Records
Industrial Facility Employee Records

Medical Records:

Type	Centralized Authority	Information System		Confidentiality Issues	Other Issues
		Computerized	Hard Copy		
Hospitals Inpatient	No	Discharge Diagnostic Indexes	Yes by Discharge Diagnosis	Yes	Variation in hospitalization Catchment patterns Quality of clinical diagnoses
Emergency Room	No	Atypical to Arrange by Disease	Yes Rarely Arranged by Disease	Yes	Variation in utilization ER Catchment patterns Quality of clinical diagnosis
Physician /Clinics	No	Atypical to Arrange by Disease	Yes Rarely Arranged by Disease	Yes	Variation in utilization Catchment patterns Quality of clinical diagnosis
School Nurses Records	Rare	Rare	Yes Rarely Arranged by Disease	Yes	Variation in utilization Quality of diagnosis
Industrial Facility Employee Records	No	Rare Atypical to Arrange by Disease	Yes Rarely Arranged by Disease	Yes	Variation in utilization Quality of diagnosis

MEDICAL RECORDS MAJOR USE:

- Allow for analysis of disease patterns in geographic areas

Advantages:

- Analysis of diseases that are not target of previously described systems

Disadvantages:

- Typically uneconomical and inefficient unless few providers in an area
- Authorization required
- Difficult to define catchment population as utilization patterns not residentially bound
- Utilization of medical care influenced by
  - disease severity
  - insurance coverage
  - accessibility
  - physician practices
  - personal preferences
- Quality of clinical diagnosis variable
- Demographic characteristics hard to obtain
- Lack of available comparative data

MEDICAL RECORDS
NATIONAL DATA SYSTEMS

NCHS National Hospital Discharge Survey
(1970-)

Summary data on sample of U.S. hospital discharges

- Diagnoses, surgical procedures, patient characteristics

426 hospitals in sample, covering >200,000 discharges/yr

Commission on Professional and Hospital Activities (CPHA):

About 40% of U.S. hospitals utilize service for compiling data on patient discharge diagnoses and characteristics. Data for a 1% representative sample of U.S. hospitals also reported (1980-).

NCHS National Ambulatory Medical Care Survey:

A sample of 3000 private physicians who voluntarily report on diagnoses and characteristics of patients office visit during a one-week period.

CENSUS DATA

Provides information on the number of persons and demographic characteristics of persons residing in geographic areas.

Major sources of census data:

1. U.S. Bureau of Census - conducts total enumeration of U.S. population by age, sex, residence, race/ethnicity, education, occupation status every 10 years.

   Census data are available in publications and in computerized format through various repositories. State health departments and some local agencies maintain census data for their areas.

   Census data can be disaggregated at various levels: national, regional, states, counties, cities, census tracts, and blocks. Disaggregation for smaller subdivisions such as census tracts or blocks may not be possible for rural areas.

The Bureau of the Census has also devised a system that is linked to latitude and longitude crosspoints. These grids can be aggregated into special areas of interest which do not strictly comply with census units.

2. Population estimates between national censuses are also prepared and published. Some estimations are based on analysis of patterns such as housing starts, utilities, tax returns, school registration, etc. Other methods are also employed. Intercensal estimates obviously are less reliable than the U. S. Census counts, may be limited to certain areas and may not hold for small subdivisions within larger areas. Such estimates are typically available from state health agencies, governmental planning agencies, etc.

Demographic information, such as locations of households, population size, and age/sex/socioeconomic/ethnic characteristics of inhabitants, is necessary to identify, enumerate, and characterize populations that may be exposed to hazardous contaminants. The primary source of demographic information of the United States population (including that of Puerto Rico and outlying areas under U.S. jurisdiction) is the Bureau of the Census of the Department of Commerce. Although the Bureau of the Census serves as the largest single source of demographic information, health assessors should not limit their search to only that information provided by the Bureau. State, regional, and local planning departments also serve as important sources of site-specific demographic data. Planning departments monitor demographic trends and can provide updated population estimates and identify areas with demographic features of special interest to the health assessor (e.g., the locations of schools, hospitals, nursing homes, minority neighborhoods, and recreational areas). Planning departments also have site-specific land-use and zoning information that is useful in the health assessment process.

The Bureau of the Census performs a United States population census once every 10 years collecting information on, among other things, population size; age, sex, and ethnic characteristics, and distribution of the population. The most recent population census was taken in 1990, and data should be available in March 1991. Collected data are organized by geographic regions (West, Midwest, South, Northeast) of the United States, and each geographic region is split into two or three divisions. Those divisions, which are groupings of states, are broken down further into a series of progressively smaller units. Some of the geographic units are governmentally defined (e.g., states and counties), and some are defined specifically for statistical purposes of the census (e.g., divisions, census tracts, and block groups). It is important to have an understanding of the geographical breakdown of the census in order to be able to properly interpret population information from the census. Relationships between the geographical units used in the census are presented in the Bureau's publication, *Census '90 Basics*. This publication also includes information about data collection and reporting for the 1990 census.

Useful demographic information can be obtained from several publications, data tapes, and maps from the Bureau. Particularly useful Bureau publications, with data for enumerating and characterizing populations near hazardous waste sites, are: *Number of Inhabitants* (Series PC80-A) and *General Population Characteristics* (Series PC80-1-B), *Census Tracts* (Series PHC80-2), and *Block Statistics* (Series PHC80-1). *Number of Inhabitants* provides population count data down to the level of incorporated and unincorporated (of greater than 1,000 inhabitants) towns and townships. The publication is available by state in either paperbound or microfiche form. *General Population Characteristics* contains data on the age, median age, sex, males per 100 females, race, and urban or rural status in geographical units down to the level of small towns (1,000 or more inhabitants). The publication is also available by state in either paperbound or microfiche form. Both publications contain relevant maps which show boundaries of the state, counties, minor civil divisions (MCDs), or census county divisions (CCDs), and all places recognized in the census. The Bureau of Census publication *Census Tracts* may also be helpful in enumerating and characterizing potentially exposed populations. This publication presents population count, age, race, and sex data for census tracts (statistical subdivisions of counties) in metropolitan statistical areas (MSAs, equaling one or more counties around a central city or urbanized area of 50,000 or more inhabitants). The usefulness of that publication may be limited, however, because many of the hazardous waste sites evaluated may not be within MSAs (1). The publication is available in paperbound, microfiche, and computer tape form by MSAs. If further resolution of population data for areas enumerated by blocks is desired, the publication *Block Statistics* (Series PHC80-1) may be consulted. That publication contains information on population statistics by block. The *Block Statistics* report series is available by state in microfiche form and is supplied with state block and MSA block index maps.

The Bureau of the Census also has several informative computer tapes available. Those data tapes are entitled *Master Area Reference Files* (MARF). The MARF tapes, referred to as MARF 1, provide basic census counts arranged from the state to block groups (BGs, subdivisions of census tracts) or enumeration districts (EDs, for areas that are not block numbered). The tapes contain numeric codes and names of geographic areas used in the 1980 Census and are available by state. MARF 2 contains, in addition to the basic census counts as in MARF 1, geographical coordinates (i.e., latitude and longitude) of population centroids. This set of tapes contains latitudinal and longitudinal coordinates and numeric codes for geographical areas and is also available by state. MARF 5, the *Zip Code Equivalency File*, associates zip codes with BGs or EDs. This set of computer tapes, available by state, contains population count data by state down to BGs or EDs and zip code areas.

Printed publications of the Bureau of the Census are available through the Government Printing Office [GPO; Superintendent of Documents, Government Printing Office, Washington, DC 20402; tel. (202) 783-3238]. Publications on microfiche or on tape can be ordered from Customer Services, Data User Services Division, Bureau of the Census, Department of Commerce, Washington, DC 20233; tel. (301) 763-4100. Some Bureau of the Census reports are accessible through other sources such as Bureau depository libraries or district offices of the Department of Commerce (refer to Appendix C of USDOC, 1986). Regional Information Services of the Bureau of the Census may be of assistance in providing information on Bureau publications. For other publications of the Bureau of the Census that may be of use in enumerating and characterizing potentially exposed populations, consult the *Census Catalog and Guide 1989* (2).

J.1. REFERENCES

1. EPA Office of Emergency and Remedial Response. *Superfund exposure assessment manual.* Washington, DC: Environmental Protection Agency, April 1988; EPA/540/1-88/001.

2. U.S. Department of Commerce. *Census catalog and guide 1989.* Washington, DC: U.S. Bureau of the Census, 1989.

3. U.S. Department of Commerce. *Census '90 Basics.* Washington, DC: U.S Bureau of the Census, Jan 1990.

4. U.S. Department of Commerce. *1990 Census of Population and Housing, Tablulation and Publication Program.* Washington DC: U.S: Bureau of the Census, July 1989.

The following information is reprinted from *Census '90 Basics*, a publication of the U.S. Department of Commerce, Bureau of the Census.

U.S. BUREAU OF THE CENSUS

The Census Bureau collects information about the Nation's people and its institutions, producing some 2,000 reports each year. Information provided by the Census Bureau is used widely in all walks of public and private life—business and industry, government, academic institutions, and community organizations—wherever there is a need for better understanding about how our society functions.

Census data form the basis for political representation and serve as benchmarks in measuring the Nation's economic well-being. Business people, community leaders, and others use Census Bureau data to study area characteristics, select sites for new facilities, assess economic potential, and so forth.

PERSPECTIVES

Historic

The 1990 Census of Population and Housing is this Nation's bicentennial census—the 21st consecutive enumeration. Since 1790, when Thomas Jefferson supervised our country's first enumeration, a population portrait has regularly marked each decade. Other statistics programs were added over the years to meet the Nation's needs. For example, the first census of manufactures was collected in 1810. Housing data entered the picture on a large scale in 1940 with the first housing census.

Contemporary

Population and housing, the main subjects of the decennial census, also are covered in surveys between censuses. Population surveys provide current estimates of population characteristics, such as the number of persons with a college education, with incomes below the poverty level, and by marital status. The Census Bureau also prepares estimates of the population by age, race, and sex, and projections of future population for the United States and individual States. It conducts special local censuses sponsored and paid for by State and local governments.

Current housing surveys furnish data comparable to those of the decennial census (but with considerably less geographic detail) as well as many other data items, such as housing inventory change, indicators of housing and neighborhood quality, and rental of new apartments.

Every 5 years, for years ending in "2" and "7," the Census Bureau conducts censuses of governments and economic activities. The census of governments shows organization, employment, and finances for State and local governments. Annual surveys keep the information current.

The Census Bureau focuses on economic activities in the censuses of agriculture, wholesale and retail trade, service industries, construction industries, manufactures, mineral industries, and transportation. To keep pace with new developments, the Census Bureau also publishes results from a variety of monthly, quarterly, and annual surveys on areas of economic activity. In addition, the Census Bureau compiles annual statistics on foreign trade and on other countries. No wonder the Census Bureau has been called "Factfinder for the Nation."

CONTENTS

CENSUS '90 Basics

THE IMPORTANCE OF THE CENSUS

Two hundred years of census-taking in America certainly qualifies the decennial count as a national tradition. The results of the enumeration of people and their housing units every 10 years reflect the social and economic fabric of American life. Accurate counts from the 1990 census will be the basis for planning our communities, making business decisions, and accomplishing a host of other activities in the last decade of this century. The census also will help provide a solid statistical framework for the 21st century.

The Census Bureau expects to employ an army of 480,000 census workers—more than the number of people who join the Armed Forces in a year—over the period 1988-91. They compile and check address lists and gather and process vital information on approximately 250 million people and 106 million housing units in the United States. They also enumerate people and housing units in Puerto Rico, the Virgin Islands of the United States, Guam, the Commonwealth of the Northern Mariana Islands, American Samoa, and Palau.

The results will be a Census Day snapshot of population, socioeconomic, and housing characteristics. The data collected will influence:

- *Political power*—The number of seats to which each State is entitled in the U.S. House of Representatives is determined by census numbers, and the boundaries selected for congressional and State legislative districts are strongly influenced by them. At the county and municipal levels, election districts must be drawn using census statistics to ensure equal representation.

- *Federal and State program funds*—Census data are used by Federal and State governments to distribute billions of dollars each year to the Nation's local governments for a wide variety of public purposes.

- *Planning for the future*—Census data are vital in planning community, private, and public facilities and services, such as shopping centers, schools, and home health care. Accurate census information is essential to help ensure the success of these developments.

ACTIVITIES FOR CENSUS '90

In 1984, the 10-year census cycle began with a year of planning and public comment—local public meetings, user conferences, Federal data forums, etc.—followed by 3 years of conducting test censuses to evaluate methods and procedures for the big count. A full dress rehearsal was held in the spring of 1988 in parts of Missouri and Washington.

A variety of activities concerned with preparing and conducting the census then followed. Here are highlights about some of them:

Checking addresses for '90 census

Compiling the Address List—The Census Bureau hired 35,000 temporary employees during 1988-89 to go door-to-door compiling a list of about 43 million addresses of housing units, many outside metropolitan areas. In addition, the Census Bureau purchased about 55 million residential addresses in large metropolitan areas from commercial mailing list companies. Census and Postal Service workers checked and updated the address list before the Census Bureau produced mailing labels for the questionnaire envelopes.

The Mailing—On March 23, 1990, the Census Bureau will mail census questionnaires to most housing units in the country; in some rural areas, the forms will be delivered

Postal workers sort census questionnaires

by census takers. Respondents will be asked to answer census questions and return forms by Census Day, April 1. Thus, the census is truly a "do-it-yourself" count. In the more sparsely settled areas and in Puerto Rico, enumerators will pick up and, if necessary, help respondents complete the questionnaires. On the night of March 20, enumerators will count people living in pre-identified shelters, on the streets, and in similar situations.

The Follow-Up—The Census Bureau will have about 300,000 temporary workers during this peak period. Many help by following up, either by phone or personal visit, at housing units for which the Census Bureau has not received a questionnaire or received one not completely filled out. Also, the enumerators collect information about vacant units and for people living in group quarters.

Local Review—The Census Bureau's Local Review Program gives local officials opportunities to point out areas where housing-unit coverage appears to be incomplete

and provide documentation to support their conclusions. The program consists of two stages: a precensus review of figures on the number of housing units derived from address lists in areas for which the Census Bureau compiled such lists, and a postcensus review of preliminary housing-unit counts from the census. Census staff will use the results of each review to pinpoint areas in need of corrective action.

Data Processing—Respondents will return most census questionnaires by mail to one of over 450 district offices (DOs) or to one of seven

Census Day is on April 1st

processing centers. The district offices will perform certain clerical checking and editing operations on the questionnaires as they are received.

After the initial processing, the questionnaires received by the district offices are packed into trucks, which are sealed for security, and sent to one of the processing centers to be worked on further.

The processing centers:

- Receive, sort, and microfilm questionnaires

- Prepare the written responses for automatic coding

- Convert microfilm into computer-readable form using FOSDIC (Film Optical Sensing Device for Input to Computers) equipment

- Edit the data by computer to check for completeness and consistency

- Transmit data electronically to Census Bureau headquarters, or send questionnaires back to district offices if additional follow-up is required.

Once the data have arrived at Census Bureau headquarters, Census Bureau staff:

- Perform computer coding, editing, and related operations to prepare the tapes (called edited detail files) with the records for all housing units and individuals for processing

- Prepare tabulations from edited detail files

- Review the resulting statistics

- Prepare data products, such as reports, computer tapes, microfiche, and compact disc—read only memory (CD-ROM) laser disks

Reporting Results—The Department of Commerce will report 1990 census counts to the President and to the States by the deadlines set forth in Title 13 of the U.S. Code, Section 141 (b)—December 31,

1990, for the total population count by State and April 1, 1991, for detailed population counts within each State. If no decision on adjustment has been made by the statutory deadlines, the results of the traditional enumeration will be published with a notation stating that these population counts are subject to possible correction for undercount or overcount. If a decision is made to adjust, the adjusted figures will be released no later than July 15, 1991. (Additional information on data products presenting 1990 census results is furnished later in the "Data Products" section.)

CONFIDENTIALITY

Census officials are highly aware of the importance of confidentiality in taking the census. Conducting a census in a democracy that values personal privacy requires special steps to gain cooperation. Title 13 of the U.S. Code authorizes the census, outlines its timing and scope (and the scope of other Census Bureau censuses and surveys), requires the public to answer the questions, mandates that all individual responses be held confidential, and sets the penalties for disclosing confidential information.

Census publications and other products generally contain combined statistics for geographic

Poster done by Jennifer Costello, a pupil in the 5th grade at the Shaw Visual and Performing Arts Center, St. Louis, Missouri

areas. Also, samples of responses (with names, addresses, and other identifying information removed) are available for users who want to design their own tabulations. The census questionnaire does not ask for Social Security numbers.

The Bureau works hard to ensure that confidentiality is not breached. Edits are performed on all data products to make sure confidential information is not released for any individual or household. Confidentiality means that only sworn employees of the Census Bureau may have access to individual census information for a period of 72 years, with the exception that individuals or their legal representatives can obtain official transcripts of information about themselves from a census for use as evidence of age, relationship, citizenship, or the like.

After 72 years, the records become public. Copies of census schedules (forms on which enumerators recorded information) from 1790 through 1910 are available, usually on microfilm, for research at the National Archives and at libraries in various parts of the country.

CENSUS CONTENT AND SAMPLE DESIGN

Information from the 1990 census will be derived either from questions asked of the entire population or from questions asked of only a sample of the population. Those questions asked about every person and housing unit are called 100-percent or short-form questions. The others are called sample or long-form questions.

Those households receiving the short-form questionnaire will be asked only the 100-percent questions, and those receiving the long-form questionnaire will be asked both the sample questions and the 100-percent questions.

Some 17.7 million housing units will receive a long form, out of an estimated total of 106 million units. Sampling rates will vary depending on geographic location and population size. Key elements of the scheme are as follows:

* Housing units in governmental jurisdictions, such as counties and incorporated places, with an estimated population of fewer than 2,500 in 1988 will be sampled at the rate of 1 in 2.

A poster on display in Census Bureau buildings

- Jurisdictions having an estimated 1988 population of 2,500 or more will be sampled at a 1-in-6 rate, except for very populous census tracts and block numbering areas (based on pre-census housing unit counts) that will be sampled at 1 in 8.

Data items that will be collected are shown in figure 1. The 1990 questions are similar to those asked in the 1980 census. This is primarily because of the continuing importance of basic facts about the population and housing of the Nation and the need to have comparable data for assessing changes occurring over the decade.

Tabulations of data from the 100-percent questions will be prepared for areas as small as a block (see description below), as well as larger areas. Because 100-percent data are not subject to sampling variability, they are accurate for areas as small as blocks.

Tabulations from the sample questions will be prepared for areas as small as block groups and for all governmental units, census tracts, and block numbering areas. The higher sample rate for small areas, described above, is expected to produce data as reliable as that for larger areas.

Processing sample data, which often involves written-in responses, will take longer than processing of 100-percent data. Therefore, 100-percent data for any area will be available before the sample data for that area. The amount of detail published will, in general, be greater for large areas like counties, large cities, metropolitan statistical areas, and States.

Figure 1.   1990 CENSUS CONTENT

100-PERCENT COMPONENT

Population
Household relationship
Sex
Race
Age
Marital status
Hispanic origin

Housing
Number of units in structure
Number of rooms in unit
Tenure—owned or rented
Value of home or monthly rent
Congregate housing (meals included in rent)
Vacancy characteristics

SAMPLE COMPONENT

Population

Social characteristics:
Education—enrollment and attainment
Place of birth, citizenship, and year of entry to U.S.
Ancestry
Language spoken at home
Migration (residence in 1985)
Disability
Fertility
Veteran status

Economic characteristics:
Labor force
Occupation, industry, and class of worker
Place of work and journey to work
Work experience in 1989
Income in 1989
Year last worked

Housing

Year moved into residence
Number of bedrooms
Plumbing and kitchen facilities
Telephone in unit
Vehicles available
Heating fuel
Source of water and method of sewage disposal
Year structure built
Condominium status
Farm residence
Shelter costs, including utilities

NOTE: Questions dealing with the subjects covered in the 100-percent component will be asked of all persons and housing units. Those covered by the sample component will be asked of a portion or sample of the population and housing units.

GEOGRAPHIC AREAS

Census data are provided for various political and statistical areas. Many are illustrated in figure 2. (See page 6.)

Political areas include:

- United States

- States, the District of Columbia, Puerto Rico, the Virgin Islands of the United States, Guam, the Commonwealth of the Northern Mariana Islands, American Samoa, and Palau

- Congressional districts

- Voting districts

- Counties

- Minor civil divisions (MCDs: legal subdivisions of counties, called townships in many States)

- Incorporated places (cities, villages, and so forth)

- American Indian reservations and associated trust lands

- Alaska Native Regional Corporations (ANRCs)

Statistical areas include:

- *Census regions and divisions*—The 50 States and the District of Columbia have been grouped into four regions, each containing two or three divisions.

- *Metropolitan statistical areas (MSAs), formerly known as standard metropolitan statistical areas (SMSAs)*—Areas consisting of one or more counties (minor civil divisions in New England) including a large population nucleus and nearby communities that have a high degree of interaction. Primary metropolitan statistical areas (PMSAs) are MSAs that make up consolidated metropolitan statistical areas (CMSAs).

- *Urbanized areas (UAs)*—Defined by population density, each includes a central city and the surrounding closely settled urban fringe (suburbs) that

Excerpt from a map showing census tracts 4050 and 4051 and their blocks in Indianapolis. (This map has been reduced.)

together have a population of 50,000 or more with a population density generally exceeding 1,000 people per square mile.

- *Urban/rural*—All persons living in urbanized areas and in places of 2,500 or more population outside of UAs constitute the "urban" population; all others constitute the "rural" population.

- *Census county divisions (CCDs)*—Statistical subdivisions of a county defined by the Census Bureau in cooperation with State officials in 21 States where minor civil divisions do not exist or are not adequate for producing subcounty statistics.

- *Census designated places (CDPs)*—Densely settled population centers without legally defined corporate limits or corporate powers.

- *Census tracts*—Small, locally defined statistical areas in metropolitan areas and some other counties. They generally have stable boundaries and an average population of 4,000.

- *Block numbering areas (BNAs)*—Areas defined, with State assistance, for grouping and numbering blocks and reporting statistics in counties without census tracts.

- *Block groups*—Groupings of census blocks within census tracts and BNAs. (These replace the enumeration districts (EDs) for which the Census Bureau provided data for many areas of the Nation in the 1980 census.)

- *Blocks*—The smallest census geographic areas, normally bounded by streets and other prominent physical features. County, MCD, and place limits also serve as block boundaries. Blocks may be as small as a typical city block bounded by four streets or as large as several square miles in rural areas. The 1990 census will be the first census in which data will be available by block for the entire Nation.

- *Alaska Native village statistical areas (ANVSAs)*—A 1990 census statistical area that delineates the settled area of each Alaska Native village (ANV). Officials of Alaska Native Regional Corporations (business and nonprofit corporate entities) outlined the ANVSAs for the Census Bureau for the sole purpose of presenting 1990 census data.

- *Tribal designated statistical areas (TDSAs)*—Geographic areas outlined for 1990 census tabulation purposes by American Indian tribal officials of recognized tribes that do not have a recognized land area.

- *Tribal jurisdiction statistical areas (TJSAs)*—Geographic areas delineated by tribal officials in Oklahoma for 1990 census tabulation purposes.

Figure 2.

Geographic Subdivisions in a Metropolitan County ....

AREA

Metropolitan Statistical Area (MSA) and Component Areas (central city and the surrounding metropolitan county(s); the Altoona, PA, MSA has only one county — "Blair" — part of which is shown here)

Census Designated Place

Incorporated Place (central city)

Urbanized Area (all shaded areas)

Incorporated Place

Minor Civil Division (MCD) or Census County Division (CCD)

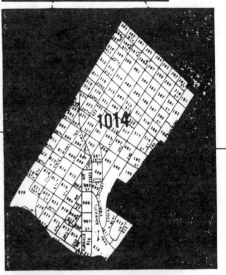

Census Tract (small, homogeneous, relatively permanent area; MSA's are subdivided into census tracts)

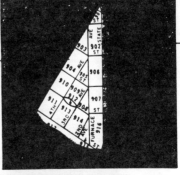

Block Group (BG; subdivision of census tracts or block numbering areas)

Block (identified throughout the country; always identified with a 3-digit number, and some have an alphabetic suffix)

.... and in a Non-Metropolitan County

POPULATION SIZE

AREA

- - These areas vary greatly
in population size

County

Place

Minor Civil Division (MCD) or
Census County Division (CCD)

Avg. pop. under
4,000

Block Numbering Area (used in
counties that do not have
census tracts)

- Average 4,000

- Average 1,000

Block Group (BG; subdivision of
census tracts or block numbering
areas; BG's in rural areas usually
are larger in area than those in more
densely settled areas)

Average 30

Average 85

Block (identified throughout
the country; blocks in rural areas
usually are larger in area than
those in more densely settled areas)

DATA PRODUCTS

The products from the 1990 census will be available in a variety of new and traditional media. The Census Bureau expects the traditional media forms—printed reports, maps, and computer tapes—to provide the largest amount of 1990 census data to users. It also will release a number of products on microfiche and CD-ROM laser disks and through its online service, CENDATA™. See the lists of products, figures 3, 5, and 6.

Printed Reports—Printed reports will be the most convenient and readily available source of data for many census users. Data presented in the 1990 census reports will be similar in kind and quantity to the data contained in reports resulting from the 1980 census. As in 1980, the reports will not include data for blocks or block groups.

The census data contained in printed reports are arranged in tables, as illustrated in figure 4. Population and housing characteristics are presented for specified geographic areas; for example, a table may present the number of rented housing units in a census tract, the number of persons 65 years of age or older in a city, or the total population of a county.

The Census Bureau will release 1990 census reports in several series, described in figure 3. Report series that present data at the small-area level, such as census tracts, will contain limited subject-matter detail (for example, counts of people by age ranges—under 5 years, 5 to 9 years, etc.—rather than by single years). Reports that include greater amounts of subject-matter detail will include less geographic detail.

There are several important differences between 1990 and 1980 reports:

- Speedier release of data. One of the Census Bureau's major goals for 1990 is quicker release of data products. Reports will be published over the period 1991 through 1993.

- Fewer report series for '90. In 1980 users had to grapple with several report series that were superseded by later reports. There will be no preliminary or advance reports for 1990.

- No reports like the *1980 Detailed Population Characteristics* or the *1980 Metropolitan Housing Characteristics*. Instead, much of the information previously shown in these reports will appear in the series of subject reports and associated computer files.

- More subject reports for 1990. Generally they will offer only national level data; some reports may include data for other

highly populated geographic areas such as States, MSAs, counties, or large cities.

- Other changes for '90. The Census Bureau is making some changes in the way that reports display race and Hispanic statistics. In 1980, a single table was repeated for each race and for Hispanics. In 1990, most reports will group together all tables for a specific race or for Hispanics, making it possible for the user to locate all the information for each group in one place.

Computer Tapes—Decennial census data have been available on computer tapes since the 1960 census. The Census Bureau provides much more data on tape than in printed reports, and all of the tabulated figures, whether in print or not, appear on computer tapes.

Uses of Computer Tapes —
- For those needing 1990 census statistics in greater subject matter or geographic detail than will be available in the printed reports.
- For those users who will need to manipulate, aggregate, or otherwise extensively process census data.
- For those who need computer files that provide codes for a wide range of geographic areas.

Public Law 94-171 Counts—These are counts that States use in legislative redistricting. This data file will be the earliest 1990 census product to provide data for areas smaller than States and the first on computer tape. These counts also will be available on CD-ROM and in printouts of the computer tape. Excerpts will be available on CENDATA. The counts will include totals for population, race groups, Hispanics, and—new to this product in 1990—population 18 years and over and housing unit counts. This means that population

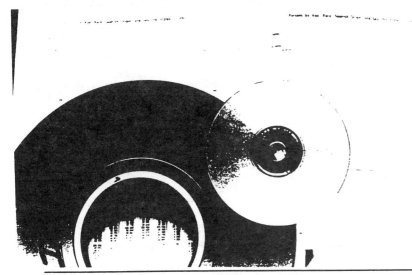

Figure 3. 1990 Census of Population and Housing Printed Reports

Series	Title	Unit of issue/ projected release date	Description	Geographic areas
		POPULATION AND HOUSING 100-percent data		
1990 CPH-1	Summary Population and Housing Characteristics	A report for the U.S./ 1992; each State and DC/1991-1992*	Population and housing unit counts, and summary statistics on age, sex, race, Hispanic origin, household relationship, units in structure, value and rent, number of rooms, tenure, and vacancy characteristics	Local governmental units (i.e., counties, incorporated places, and towns and townships) and American Indian and Alaska Native areas
1990 CHP-2	Population and Housing Unit Counts	A report for the U.S./ 1992; each State and DC/1991-1992*	Total population and housing unit counts for 1990 and previous censuses	States, counties, minor civil divisions (MCDs)/ census county divisions (CCDs), places, and summary geographic areas (for example, urban and rural, metropolitan and nonmetropolitan residence)
		100-percent and sample data		
1990 CPH-3	Population and Housing Characteristics for Census Tracts and Block Numbering Areas	A report for each MSA and each State/1992-1993*	Statistics on 100-percent and sample population and housing subjects	In MSAs: census tracts/ BNAs, places of 10,000 or more inhabitants, and counties. In the remainder of each State: census tracts/BNAs, places if 10,000 or more, and counties
1990 CPH-4	Population and Housing Characteristics for Congressional Districts of the 103rd Congress	A report for each State and DC/1993	Statistics on 100-percent and sample population and housing subjects	Congressional districts (CDs) and within CDs, counties, places of 10,000 or more inhabitants, MCDs of 10,000 or more in selected States, and American Indian and Alaska Native areas
		Sample data		
1990 CPH-5	Summary Social, Economic, and Housing Characteristics	A report for the U.S./1993; each State and DC/1992*	Statistics on 100-percent and sample population and housing subjects	Local governmental units (i.e., counties, incorporated places, and towns and townships in selected States) and American Indian and Alaska Native areas

*Reports for Puerto Rico and the Virgin Islands of the United States are included.

Figure 3. 1990 Census of Population and Housing Printed Reports—Con.

Series	Title	Unit of issue/ projected release date	Description	Geographic areas
		POPULATION 100-percent data		
1990 CP-1	General Population Characteristics	A report for the U.S./1992; each State and DC/1992*	Detailed statistics on age, sex, race, Hispanic origin, marital status, and household relationship characteristics	States, counties, places of 1,000 or more inhabitants, MCDs of 1,000 or more in selected States, State parts of American Indian and Alaska Native areas, and summary geographic areas such as urban and rural
1990 CP-1-1A	General Population Characteristics for American Indian and Alaska Native Areas	A single report/ 1992	Detailed statistics on age, sex, race, Hispanic origin, marital status, and household relationship characteristics	American Indian and Alaska Native areas, i.e., American Indian reservations, trust lands, tribal jurisdiction statistical areas in Oklahoma, Alaska Native village statistical areas and Alaska Native regional corporations
1990 CP-1-1B	General Population Characteristics for Metropolitan Statistical Areas	A single report/ 1992	Detailed statistics on age, sex, race, Hispanic origin, marital status, and household relationship characteristics	Individual MSAs and their component areas. For MSAs split by State boundaries, summaries are provided both for the parts and for the whole MSA
1990 CP-1-1C	General Population Characteristics for Urbanized Areas	A single report/ 1992	Detailed statistics on age, sex, race, Hispanic origin, marital status, and household relationship characteristics	Individual UAs and their component areas. For UAs split by State boundaries, summaries are provided both for the parts and for the whole UA
		Sample data		
1990 CP-2	Social and Economic Characteristics	A report for the U.S., each State and DC/1993*	Statistics on 100-percent and sample population subjects	States (including summaries such as urban and rural), counties, places of 2,500 or more inhabitants, MCDs of 2,500 or more in selected States, and the State portion of American Indian and Alaska Native areas

*Reports for Puerto Rico and the Virgin Islands of the United States are included.

Figure 3. 1990 Census of Population and Housing Printed Reports—Con.

Series	Title	Unit of issue/ projected release date	Description	Geographic areas
		POPULATION—Con. Sample data—Con.		
1990 CP-2-1A	Social and Economic Characteristics for American Indian and Alaska Native Areas	A single report/ 1993	Statistics on 100-percent and sample population subjects	American Indian and Alaska Native areas. as for CP-1-1A
1990 CP-2-1B	Social and Economic Characteristics for Metropolitan Statistical Areas	A single report/ 1993	Statistics on 100-percent and sample population subjects	Individual MSAs and their component areas. as for CP-1-1B
1990 CP-2-1C	Social and Economic Characteristics for Urbanized Areas	A single report/ 1993	Statistics on 100-percent and sample population subjects	Individual UAs and their component areas, as for CP-1-1C
1990 CP-3	Population Subject Reports	A report for each subject/1993	Approximately 30 reports on population census subjects such as migration, education, income, the older population, and racial and ethnic groups	Generally limited to the U.S., regions, and divisions; for some reports, other highly populated areas such as States, MSAs, counties, and large cities
		HOUSING 100-percent data		
1990 CH-1	General Housing Characteristics	A report for the U.S./ 1992; each State and DC/1992*	Detailed statistics on units in structure, value and rent, number of rooms, tenure, and vacancy characteristics	States, counties, places of 1,000 or more inhabitants. MCDs of 1,000 or more in selected States, State parts of American Indian and Alaska Native areas, and summary geographic areas such as urban and rural
1990 CH-1-1A	General Housing Characteristics for American Indian and Alaska Native Areas	A single report/ 1992	Detailed statistics on units in structure, value and rent, number of rooms, tenure, and vacancy characteristics	American Indian and Alaska Native areas, i.e., American Indian reservations, trust lands, tribal jurisdiction statistical areas in Oklahoma, Alaska Native village statistical areas, and Alaska Native regional corporations

*Reports for Puerto Rico and the Virgin Islands of the United States are included.

Figure 3.   1990 Census of Population and Housing Printed Reports—Con.

Series	Title	Unit of issue/ projected release date	Description	Geographic areas
		HOUSING—Con. 100-percent data—Con.		
1990 CH-1-1B	General Housing Characteristics for Metropolitan Statistical Areas	A single report/ 1992	Detailed statistics on units in structure, value and rent, number of rooms, tenure, and vacancy characteristics	Individual MSAs and their component areas. For MSAs split by State boundaries, summaries are provided both for the parts and for the whole MSA
1990 CH-1-1C	General Housing Characteristics for Urbanized Areas	A single report/ 1992	Detailed statistics on units in structure, value and rent, number of rooms, tenure, and vacancy characteristics	Individual UAs and their component areas. For UAs split by State boundaries, summaries are provided both for the parts and for the whole UA
		Sample data		
1990 CH-2	Detailed Housing Characteristics	A report for the U.S./1993; each State and DC/1992-1993*	Statistics on 100-percent and sample housing subjects	States (including summaries such as urban and rural), counties, places of 2,500 or more inhabitants, MCDs of 2,500 or more in selected States, State parts of American Indian and Alaska Native areas
1990 CH-2-1A	Detailed Housing Characteristics for American Indian And Alaska Native Areas	A single report/ 1993	Statistics on 100-percent and sample housing subjects	American Indian and Alaska Native Areas, as in CH-1-1A
1990 CH-2-1B	Detailed Housing Characteristics for Metropolitan Statistical Areas	A single report/ 1993	Statistics on 100-percent and sample housing subjects	Individual MSAs and their component areas, as in CH-1-1B
1990 CH-2-1C	Detailed Housing Characteristics for Urbanized Areas	A single report/ 1993	Statistics on 100-percent and sample housing subjects	Individual UAs and their component areas, as in CH-1-C
1990 CH-3	Housing Subject Reports	A report for each subject/1993	Approximately 10 reports on housing census subjects such as structural characteristics and space utilization	Generally limited to U.S. regions, and divisions; for some reports, other highly populated areas such as States, MSAs counties, and large cities

*Reports for Puerto Rico and the Virgin Islands of the United States are included.

Figure 4. Table From a 1980 Census Report—PHC80-3, *Summary Characteristics for Governmental Units and Standard Metropolitan Statistical Areas*

Table 1. **Summary of General Population Characteristics: 1980**

For meaning of symbols see introduction. For definitions of terms, see appendixes A and B]

The State Standard Metropolitan Statistical Areas Counties and County Subdivisions Incorporated Places	Total	Fe-male	Under 5 years	8 years and over	65 years and over	Median age	White	Black	American Indian Eskimo and Aleut	Asian and Pacific Islander	Spanish origin	In group quarters	House-holds	Persons per house-hold	Families
The State	5 490 224	51 4	7 6	70 5	10 7	29 2	5 004 394	414 785	7 836	20 557	87 047	145 681	1 927 050	2 77	455 556
SMSA'S															
Anderson Ind	139 336	51 4	7 3	70 2	10 9	30 2	128 913	9 652	155	294	789	3 577	49 985	2 72	38 061
Bloomington Ind	98 785	50 7	5 6	79 0	7 2	24 6	93 472	2 570	130	1 139	1 104	15 062	33 952	2 47	2 099
Cincinnati Ohio-Ky-Ind	34 291	50 8	8 1	68 0	10 8	29 5	33 930	227	27	79	128	294	11 486	2 96	9 196
Elkhart Ind	137 330	51 7	8 3	69 1	9 6	29 0	129 971	5 761	256	491	1 719	1 973	48 148	2 81	35 944
Evansville Ind -Ky	268 559	52 1	7 3	72 6	12 7	30 9	253 722	13 149	286	863	1 171	5 346	99 020	2 60	73 132
Fort Wayne Ind	382 961	51 4	8 3	69 2	9 9	28 6	350 650	26 435	645	1 422	6 052	5 689	134 313	2 81	100 521
Gary-Hammond-East Chicago Ind	642 781	51 2	8 5	68 5	8 4	28 1	491 274	126 350	829	2 627	46 621	7 901	214 244	2 96	167 101
Indianapolis Ind	144 575	51 9	7 5	70 4	9 8	29 3	354	486	5 437	8 839	21 154	418 485	2 74	
Lafayette Ind		51 7	7 7	69 0	9 4				209	295			37 043		

and housing unit counts from the smallest areas (blocks) to the largest will be available within a year after the census, including voting districts where States have identified them for the Census Bureau.

In addition to this redistricting file, the Census Bureau also will offer summary tape files, subject files, public use microdata sample files, and special files.

Summary Tape Files (STFs)— The STF's are designed to provide statistics with greater subject detail for geographic areas than is feasible or desirable to provide in printed reports. They will contain essentially the same types of information as in the reports—just more of it.

Here are some STF highlights:

- STFs 1 and 2 will contain 100-percent data, and STFs 3 and 4 will offer sample data.
- Each STF will present a particular set of data tables for specific types of geographic areas.
- Each STF will have three or more file types (indicated by a letter suffix attached to the STF number) that differ in the types of geographic detail reported, but contain the same data detail.

- STFs 1 and 3 will have more geographic detail, but less data detail than STFs 2 and 4.
- STFs 1 through 4 will be comparable to those produced in 1980.
- STF 5, released in 1980 for the United States, each State, and the District of Columbia, will not be issued in 1990. Similar data will be provided through the 1990 subject reports and related summary tape files, noted below.

Subject Summary Tape Files— These files will correspond to the subject reports and, in some cases, contain data for States, counties, and large cities.

Public Use Microdata Sample Files (PUMS)—These microdata files on computer tape will present a sample of unidentified long-form housing-unit records for large geographic areas. Each housing-unit sample record will present essentially all the census data collected about each person in a sample household plus the housing unit's characteristics. Information that might tend to identify an individual or a housing unit will not be contained on the file.

Microdata files enable users with special needs to prepare customized tabulations and cross-tabulations of virtually any item

on the census questionnaire. There will be two sets of public-use micro-data files.

- A 5-percent sample of housing units in which each household record includes codes to let the user know in what group of counties (or sometimes a county or smaller area) the household is located.
- A file presenting a 1-percent sample of housing units. It shows data for metropolitan statistical areas (MSAs) in the 1990 census and other large areas. In 1980, there were two files with 1-percent samples rather than the one file for 1990. The 1980 1-percent sample showing data for large States and groups of smaller States will not be produced in 1990.

The minimum population size of any county group or other area identified is, as in 1980, 100,000 persons. What is new for 1990 is that household and geographic-area identifiers have been added to each person-record to make the files easier to process.

Other Special Computer Tape Files—Other files are planned, such as the Census/Equal Employment Opportunity (EEO) File and the County-to-County Migration File. See figure 6.

Figure 5. 1990 Census of Population and Housing Summary Tape Files

Summary Tape File (STF 1A, 1B, etc.) and data type (100 percent or sample)		Projected release date	Geographic areas	Description
STF 1 (100 percent)	A*†	1991-1992	States, counties, MCDs/CCDs, places, census tracts/ BNAs, block groups (BGs)	About 1,000 cells/ items of 100-percent population and housing counts and characteristics for each geographic area
	B*†	1991-1992	States, counties, MCDs/CCDs, places, census tracts/ BNAs, BGs, blocks	
	C†	1992	U.S., regions, divisions, States (including summaries such as urban and rural), counties, places of 10,000 or more inhabitants, MCDs of 10,000+ in selected States, MSAs, UAs, American Indian and Alaska Native areas	
	D	1993	Congressional districts (CDs) by State; and within each CD: counties, places of 10,000+, MCDs of 10,000+ in selected States	
STF 2 (100 percent)	A	1992	In MSAs: census tracts/BNAs, places of 10,000 or more inhabitants, and counties. In the reminder of each State: counties, places of 10,000+, and census tracts/ BNAs	Over 2,000 cells/items of 100-percent population and housing counts and characteristics for each geographic area. Each of the STF 2 files will include a set of tabulations for the total population and separate presentations of tabulations by race and Hispanic origin
	B	1992	States (including summaries such as urban and rural), counties, places of 1,000+, MCDs of 1,000+ in selected States, State portion of American Indian and Alaska Native areas	
	C	1992	U.S., regions, divisions, States (including summaries such as urban and rural), counties, places of 10,000+, MCDs of 10,000+ in selected States, MCDs of fewer than 10,000+ in New England MSAs, American Indian and Alaska Native areas, MSAs, UAs	
STF 3 (sample)	A*†	1992	States, counties, MCDs/CCDs, places, census tracts/ BNAs, BGs	Over 2,300 cells/items of sample population and housing characteristics for each geographic area
	B†	1993	5-digit ZIP Codes within each State	
	C†	1993	U.S., regions, divisions, States, counties, MCDs of 10,000+ in selected States, places of 10,000+, American Indian and Alaska Native areas, MSAs, UAs	
	D	1993	Congressional districts (CDs) by State; and within each CD: counties, places of 10,000+, MCDs of 10,000+ in selected States	
STF 4 (sample)	A	1992	In MSAs: census tracts/BNAs, places of 10,000 or more inhabitants, and counties. In the remainder of each State: counties, places of 10,000+, and census tracts/BNAs	Over 8,000 cells/items of sample population and housing characteristics for each geographic area. Each of the STF 4 files will include a set of tabulations for the total population and separate presentations of tabulations by race, Hispanic origin, and possibly selected ancestry groups
	B	1992-1993	States (including summaries such as urban and rural), counties, places of 2,500+, MCDs of 2,500+ in selected States, MCDs of fewer than 2,500 in New England MSAs, State portion of American Indian and Alaska Native areas	
	C	1993	U.S., regions, divisions, States (including urban and rural and metropolitan and nonmetropolitan components), counties, places of 10,000+, MCDs of 10,000+ in selected States, MCDs of fewer than 10,000 in New England MSAs, American Indian and Alaska Native areas, MSAs, UAs	

*Available on microfiche (STF 1B microfiche is an extract).
†Available on laser disk (CD-ROM). (STF 1B data on CD-ROM is an extract)

Figure   6. Special Computer Tape Files and Other Data Products

Title	Projected release date	Description	Geographic areas
Public Law 94-171 Counts File (redistricting data)	Prior to April 1991	Statistics on total population, age, sex, race, Hispanic origin available in paper copy and computer files; housing unit counts on computer tape files only	States, counties, MCDs/ CCDs, places, census tracts/ BNAs, BGs, and blocks; and voting districts where States have identified them for the Census Bureau
Census/Equal Employment Opportunity (EEO) Special File	1992	Sample tabulations showing detailed occupations and educational attainment data by age. Cross tabulated by sex, Hispanic origin, and race	Counties, MSAs, places of 50,000 or more inhabitants
County-to-County Migration File	1993	Summary statistics for all intrastate county-to-county migration streams and significant interstate county-to-county migration streams. Each record will include codes for the geographic area of destination, and selected characteristics of the persons who made up the migration stream	States, counties
Public Use Microdata Sample (PUMS) Files		Computerized files containing a sample of individual long-form census records showing most population and housing characteristics but with identifying information removed	
5 Percent—County Groups	1993		County groups or smaller areas with 100,000 or more inhabitants
1 Percent—Metropolitan Statistical Areas identified in 1990	1993		MSAs and other large areas with 100,000 or more
User-Defined Area Tabulations		Printouts, tapes or other products with data tables, maps, and narrative (if requested). The table format will be standard or fixed for these tabulations	User-defined areas created from blocks
Special Tabulations		Special tabulations for any specific geographic or subject-matter area in any format (reports, tapes, and microfiche)	User-defined areas or standard areas

Microfiche—Block statistics will be available on microfiche as they were in 1980. The microfiche will present a subset of the tabulations for blocks found in STF 1B. In 1990, for the first time, the entire Nation is blocked. This will increase the number of blocks for which the Census Bureau provides data from 2.5 million in 1980 to about 7 million in 1990. The cost and storage of block data of this magnitude would be prohibitive if they were published in printed reports.

STFs 1A and 3A will be made available on microfiche, as well.

Other Media—Newer media also will be used for 1990 products:

• *Online information systems*— The Census Bureau began its own online information service called CENDATA™ in 1984. A number of Census Bureau reports, in whole or in part, are offered online. For 1990, CEN-DATA will provide up-to-the-day information about the availability of data products and will carry selections of State, county, metropolitan statistical area, and place data from the Public Law (PL) 94-171 tape file and STFs 1 and 3. CENDATA is available through two information vendors, CompuServe and DIA-LOG.

• *CD-ROM*—Compact disk-read only memory, a type of optical or laser disk, is the most recently developed technology for data storage and retrieval. One 4 3/4-inch CD-ROM can hold the contents of approximately 1,500 flexible diskettes, or three or four high density tapes. For 1990, the Public Law 94-171 tape file and STFs 1A, 1B (extract), 1C, 3A, 3B, and 3C will be available on CD-ROM.

Custom Data Programs—These programs are for users who require unique tabulations that are not included in Census Bureau standard products; for example, information for locally defined geographic areas. Any data that the Census Bureau furnishes will be subject to the usual standards to ensure that confidential individual information is not revealed.

• *User-Defined Area Tabulations*— This program will provide data for locally defined geographic areas that do not correspond to standard census geographic areas. Users will identify the geographic areas of interest to them by compiling census blocks. The Census Bureau then will create a set of predefined tables of information for these areas.

• *Special Tabulations*—The Census Bureau will prepare special tabulations, paid for by the requester, for any specific geographic or subject-matter area. Standard reports, tapes, and microfiche should be used whenever possible, though, since special tabulations are fairly expensive and can only be done when the demands of regular work permit.

MAPS AND GEOGRAPHIC FILES

Maps—Census maps are necessary for virtually all uses of small-area census data. They are needed to locate specific geographic areas and study the spatial relationship of the data for analytic purposes. The Census Bureau plans to offer a variety of 1990 census maps. Among them will be these three series:

• *County block maps*—These maps will show census blocks and their numbers, other boundaries, and physical features. Maps will be available by county with one or more map sheets each, depending on the size, shape, number of blocks, and density of the block pattern.

• *County subdivision maps*— Maps in this State-based series present the boundaries of the counties, county subdivisions (MCDs or CCDs), places, American Indian reservations (including off-reservation trust lands), and Alaska Native areas.

• *Census tract/BNA outline maps*—These maps depict census tract/block numbering area (BNA) boundaries and features underlying the boundaries.

Geographic Publications—The Census Bureau will produce a Geographic Identification Code Scheme (GICS) report that shows the 1990 census geographic area codes for States, counties, county subdivisions, places, and certain other areas, along with some descriptive information.

Machine-Readable Geographic Files—All 1990 summary tape files will include 1990 census geographic area codes and other geographic information, similar to that presented on the 1980 Master Area Reference File 2.

The Census Bureau has developed an automated geographic data base, known as the TIGER (Topologically Integrated Geographic Encoding and Referencing) System, that will allow the production of various geographic products to support the 1990 census. It provides coordinate-based digital map information for the entire United States, Puerto Rico, the Virgin Islands, and the Pacific territories over which the United States has jurisdiction.

The TIGER System will significantly improve 1990 census maps and geographic reference products and will permit users to generate, by computer, maps at different scales for any geographic area of the country.

The final list of TIGER products has not been determined, but the Census Bureau expects that extracts from the TIGER System will be released in several formats. One extract of selected geographic and cartographic information is called the TIGER/Line files.

Digitizing geographic coordinates into a TIGER/Line file

TIGER/Line files contain basic data for the segments of each boundary or feature (e.g., roads, railroads, and rivers), including adjacent census geographic area codes, latitude/longitude coordinates of segment end points, the name and type of the feature, and the relevant census feature class code identifying the feature segment by category. TIGER/Line files also furnish address ranges and associated ZIP Codes for each side of street segments for areas approximating the urbanized areas.

The TIGER/Line files are organized by county and are available to the public now in a precensus version and later in a final 1990 census version. The files are released on computer tape and CD-ROM.

TAPPING OTHER CENSUS BUREAU RESOURCES

The Census Bureau has more to offer than just the results of the census of population and housing. Through other censuses, surveys, and estimates programs, it compiles and publishes data on subjects as diverse as appliance sales, neighborhood conditions, and

exports to other countries. The other censuses, such as retail trade, manufactures, and governments, are collected for years ending in "2" and "7." Surveys and estimates programs generate results as often as every month.

Here are examples of the information published about—

- *People:* Age, race, sex, income, poverty, child care, child support, fertility, noncash benefits, education, commuting habits, pension coverage, unemployment, ancestry.

- *Business and industry:* Number of employees, total payroll, sales and receipts, products manufactured or sold.

- *Housing and construction:* Value of new construction, numbers of owners and renters, property value or rent paid, housing starts, fuels used, mortgage costs.

- *Farms:* Number, acreage, livestock, crop sales.

- *Governments:* Revenues and expenditures, taxes, employment, pension funds.

- *Foreign trade:* Exports and imports, origin and destination, units shipped.

- *Other nations:* Population, birth rates, death rates, literacy, fertility.

Many of the monthly "economic indicators" that measure how the Nation is doing come directly or indirectly from the Census Bureau. Examples: employment and unemployment; housing starts; wholesale and retail trade; manufacturers' shipments, inventories, and orders; export and import trade; and sales of single-family homes.

Use the attached form to request more information about any of these subjects.

HOW TO OBTAIN REFERENCE MATERIALS AND ASSISTANCE

Key Reference Sources—The Census Bureau issues several reference publications that are of value to many data users.

- *1990 Census of Population and Housing Tabulation and Publication Program*—This free report gives more complete descriptions of 1990 products, estimated publication dates, and a comparison of 1990 products with those of 1980, and more. Order by sending in the attached form.

- *Census ABC's–Applications in Business and Community*—A free report that highlights key information about the 1990 census and illustrates a variety of ways the data can be used. Order by sending in the attached form.

- *1990 Census of Population and Housing Guide*—The primary guide to the 1990 census that will provide detailed information about all aspects of the census and a comprehensive glossary of census terms. Sign up to receive an order form for it as soon as it's ready by sending in the attached form.

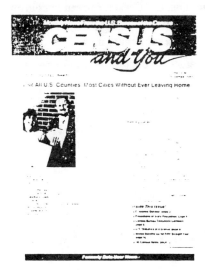

- *Census and You*—The Census Bureau's monthly newsletter for data users. It reports on the latest 1990 census developments, selected new publications and computer tapes, other censuses and surveys, developments in Census Bureau services to users, and upcoming conferences and training courses. (Use the attached form to request a sample copy.) A subscription costs $12 per year. Make checks payable to "Superintendent of Documents," and send to the Superintendent of Documents, U.S. Government Printing Office, Washington, DC 20402; or call 202/783-3238 to charge to a VISA, MasterCard, or GPO deposit account. Cite the code "DUN" in your order.
- *1990 Census Publication Order Forms*—For basic information (a

brief description, prices, and stock numbers) on ordering 1990 census reports, get on the mailing list to receive order forms as the reports are published. Complete and return the attached form. For information about all 1990 census products and products from other censuses and surveys, too, subscribe to *Monthly Product Announcement*, described next.

- *Monthly Product Announcement (MPA)*—A free monthly listing of all new Census Bureau publications; microfiche; maps; data files on tape, diskettes, or CD-ROM's; and technical documentation. 1990 census products will be specially noted. For a subscription, complete and mail in the attached form.
- *Census Catalog and Guide*—A comprehensive annual description of Census Bureau data products (from 1980 to date), statistical programs, and services of the Census Bureau. It provides abstracts of the publications, data files, microfiche, maps, and items online. In addition, the *Census Catalog and Guide* offers such features as information about censuses and surveys and telephone contact lists for data specialists at the Census Bureau, the State data centers, and other data processing service centers. The cost for the 1989 edition is $21. When ordering, give the stock number: 003-024-07009-0. (It is sold by the Superintendent of Documents; see ordering instructions for *Census and You*, above.)

Users also can get listings of new Census Bureau products, updated daily, and selected statistical information online through CENDATA™ the Census Bureau's online information service. CompuServe (800/848-8199) and DIALOG (800/334-2564), which are information service companies, offer CENDATA to their customers. For more information, complete and return the attached form.

Other Census Services—The Census Bureau has specialists whom users may consult at the Washington headquarters and its 12 regional offices. They answer inquiries by telephone, correspondence, and personal visit. From time to time, they also conduct workshops, seminars, and training courses.

- *Washington contact*—For general and product ordering information: Customer Services, Bureau of the Census, Washington, DC 20233 (301/763-4100).

- *Regional office contacts*—

Atlanta, GA	404/347-2274
Boston, MA	617/565-7078
Charlotte, NC	704/371-6144
Chicago, IL	312/353-0980
Dallas, TX	214/767-7105
Denver, CO	303/969-7750
Detroit, MI	313/354-4654
Kansas City, KS	816/891-7562
Los Angeles, CA	818/892-6674
New York, NY	212/264-4730
Philadelphia, PA	215/597-8313
Seattle, WA	206/728-5314

Other Sources of Census Products and Services—

- *State Data Centers*—The Census Bureau furnishes data products, training in data access and use, technical assistance, and consultation to all States, the District of Columbia, Guam, Puerto Rico, and the U.S. Virgin Islands. State Data Centers, in turn, offer publications for reference, printouts from computer tape, specially prepared reports, and other products and assistance to data users. For a list of the State data centers, complete and return the attached form. The list also will note organizations in the States participating in the Census Bureau's Business/Industry Data Center (BIDC) Project. The BIDCs help business people, economic development planners, and other data users obtain and use data.

- *National Clearinghouse*—The National Clearinghouse for Census Data Services is a listing of private companies and other organizations that offer assistance in obtaining and using data released by the Census Bureau. For a list of participants in the National Clearinghouse, complete and return the attached form.

- *National Services Program*—The National Services Program (NSP) provides a structure for cooperation between the Census Bureau and nationally based organizations that represent minorities or other segments of the population who have been historically undercounted in decennial censuses. The participants include social service, business, professional, civil rights, educational, and religious groups. To learn more about the NSP, write to the National Services Program, Data User Services Division, Bureau of the Census, Washington, DC 20233, or call 301/763-1384.

- *Depository Libraries*—There are 1,400 libraries that receive (from the Government Printing Office) those Federal publications which they think their patrons will need. Often some of these publications are Census Bureau reports. The Census Bureau provides free reports to an additional 120 Census depository libraries. Also, many libraries purchase census reports for their areas.

For Further Information—To receive additional information on the 1990 census as it becomes available, and on Census Bureau products in general, mail the attached form.

Census '90 Basics

Please send me more information about Census Bureau programs and products

(Check those you are interested in.)

☐ *1990 Census of Population and Housing Tabulation and Publication Program*

☐ *Census ABC's—Applications in Business and Community*

☐ *1990 Census of Population and Housing Guide order form*—as soon as it is ready

☐ Order forms for 1990 census reports as soon as they are prepared

☐ Free sample copy of *Census and You*, the Census Bureau's monthly newsletter

☐ Free subscription to *Monthly Product Announcement*

☐ Lists of State Data Centers, Business/Industry Data Centers (BIDCs), and Clearinghouse organizations

☐ Information about CENDATA, the Census Bureau's online service

Information about Census Bureau statistics on:

☐ Agriculture	☐ International
☐ Business	☐ Manufacturing
☐ Construction	☐ Mineral Industries
☐ Foreign trade	☐ Population
☐ Governments	☐ Transportation
☐ Housing	

Name:_____

Title:_____

Organization:_____

Street:_____

City:_____

State:_____ ZIP:_____

Phone: (____)-_____-_____

Detach this form and mail to:

**Customer Services
Bureau of the Census
Washington, DC 20233**

THE METRIC SYSTEM

The metric system, established by international treaty at the Metric Convention in Paris in 1875, has been extended and improved. The currently established official metric system is called the International System of Units (SI). The metric system has seven fundamental units (Table K.1.) from which all others can be

Table K.1. Metric System Units

Quantity	Unit	Symbol
length	meter	m
mass	gram	g
time	second	s, or sec
electric current	ampere	A, or amp
temperature	Kelvin	K, or $^\circ$K
luminous intensity	candela	cd
amount of substance	mole	mol

derived using the prefixes shown in Table K.2. Table K.3. identifies conversion factors that can be used to convert from English to SI units.

Table K.2. Metric System Prefixes

Prefix	Symbol	Multiplier
mega	M	1,000,000 or 10^6
kilo	k	1,000 or 10^3
deci	d	0.1 or 10^{-1}
centi	c	0.01 or 10^{-2}
milli	m	0.001 or 10^{-3}
micro	μ	0.000001 or 10^{-6}
nano	n	0.000000001 or 10^{-9}

CALCULATING ATMOSPHERIC CONCENTRATIONS

Concentrations of atmospheric contaminants are usually expressed either as the weight or mass of the contaminant per specified volume of air (e.g., mg/m^3) or as the volume of contaminant gas or vapor per specified volume of air, such as ppm. Other common measures include mg/L and μg/L. An example would be if air were sampled for 30 minutes at a sampling rate of 2 liters of air per minute, and an analysis of the sample showed that 0.5 mg of the contaminant had been collected. In this case, the concentration of contaminant, C, in terms of mg/m^3 would be:

$$C = (0.5\,mg/30\,min) \times (1\,min/2L) \times (1000\,L/1\,m^3)$$
$$C = 8.3\ mg/m^3$$

VOLUME PER UNIT VOLUME

The concentration of gases, vapors, and liquids is often expressed as volume of contaminant in a specific volume of air or water, referred to as volume per unit volume. This is most conveniently expressed as parts per million (ppm) or percent by volume. Parts per million is the volume of contaminant per million volumes of air. Any volume unit can be used as long as the units for both parts are the same (e.g., liters of contaminant per million liters of air or water). Measurements in percent volume are less applicable to hazardous waste characterization because they represent very high concentrations that would not usually be found in environmental contamination situations. Percent by volume can be thought of as parts per hundred, so:

$$(\% \text{ by volume}) \times (10,000) = ppm$$

Table K.3. U.S. to SI Conversions

U.S. Name	Abbreviation	Conversion Factor	Symbol	SI Name
acre	acre	0.045	ha	hectare
acre-foot	acre-ft	1.234	m^3	cubic meters
cubic foot	ft^3	28.32	L	liter
cubic inch	in^3	16.387	mL	milliliters
cubic foot/minute	ft^3/min	28.32	L/min	liters/minute
feet/second	ft/s	0.305	m/s	meters/second
foot	ft	0.305	m	meters
gallon	gal	3.785	L	liters
gallons/acre/day	gal/acre/d	9.353	L/ha/d	liters/hectare/day
gallons/day	gal/d	4.381E-5	L/d	liters/day
gallons/minute	gal/min	0.0631	L/s	liters/second
inches	in	2.54	cm	centimeters
inches/hour	in/h	2.54	cm/h	centimeters/hour
mile	mi	1.609	km	kilometers
miles/hour	mi/h	0.45	km/h	kilometers/hour
million gallons	Mgal	3.785	ML	megaliters
million gal/acre	Mgal/acre	8.353	m^3/ha	cubic meters/hectare
million gal/day	Mgal/day	43.0	L/s	liters/second
ounce	oz	28.35	g	grams
parts per million	ppm	1.0	mg/L	milligrams/liter
pounds	lb	0.454	kg	kilograms
pounds/acre/day	lb/acre/d	1.12	kg/ha/d	kilograms/hectare/day
pounds/cubic foot	lb/ft^3	0.0162	g/cm^3	grams/cubic centimeter
pounds/square inch	lb/in^2	0.069	kg/cm^2	kilograms/square cm
square foot	ft^2	0.0929	m^2	square meters
square inch	in^2	6.452	cm^2	square centimeters
yard	yd	0.914	m	meters

To convert from the English Unit to the SI Unit, multiply by the conversion factor. For example, to convert 4 feet to meters: 4 ft x 0.305 = 1.22 meters.

To convert from the SI Unit to the English Unit, divide by the conversion factor. For example, to convert 1.22 meters to feet: 1.22 meters / 0.305 = 4 ft.

TEMPERATURE AND PRESSURE CONVERSIONS

When concentration is expressed in terms of mass of contaminant per volume of air, changes in temperature and pressure result in changes in the volume of air but not in the mass of the contaminant. Therefore, the concentration expressions are accurate only at a specified temperature and pressure. Because of this relationship, it is necessary to understand the use of molecular weights and molecular volumes. Likewise, it is necessary to understand the effects of pressure and temperature changes and to be able to use these in calculations.

Absolute temperature must be used when calculating changes in gas volumes. Absolute temperature is based on a theoretical absolute zero point, which has been determined to be $-273^\circ C$ or $0^\circ K$. The absolute temperature scale is calculated as follows:

$$^\circ K = {}^\circ C + 273$$

where,

$^\circ K$ = temperature in Kelvin
$^\circ C$ = temperature in Centigrade

Changes in gas volume can be calculated based on the following relationship:

$$\frac{V1}{V2} = \frac{T1}{T2}$$

If, for example, a volume of gas is 40 ft^3 at 40°F, what is its volume at 70°F? First, convert from Fahrenheit to Centigrade using the following formula.

$$
\begin{aligned}
^\circ C &= 5/9 \times ({}^\circ F - 32) \\
&= 5/9 \times (40-32) \\
&= 4.4
\end{aligned}
$$

Then convert the temperature to Kelvin.

$$
\begin{aligned}
^\circ K &= {}^\circ C + 273 \\
&= 4.4 + 273 \\
&= 277.4
\end{aligned}
$$

Using the same method, 70 $^\circ C$ is equal to 294.1 $^\circ K$. Next calculate the new volume.

$$\frac{V1}{V2} = \frac{T1}{T2}$$

$$\frac{40 \text{ ft}^3}{V2} = \frac{277.4}{294.1}$$

$$V2 = 42.4 \text{ ft}^3$$

Changes in pressure usually do not affect volume to a large degree. If large changes in pressure occur, pressure must be factored into the calculation. The Table K.4. gives conversion factors for pressure based on altitude, where:

selected altitude volume = volume at 0 ft × factor

Table K.4. Altitude Pressure Conversion Factors

Altitude Above Sea Level	Factor	Altitude Above Sea Level	Factor
-1000	0.965	3000	1.115
-500	0.982	4000	1.156
0	1.0	5000	1.200
500	1.018	6000	1.248
1000	1.035	7000	1.295
2000	1.074	8000	1.349

It is important to note that a volume of gas is assumed to be at standard temperature (0° C) and pressure (760 mm Hg or 1 atmosphere) unless otherwise stated. At times, "normal" temperature and pressure are given referring to 25°C and 760 mm Hg. To correctly calculate concentrations of a contaminant in air, use the following equation:

$$ppm = C \times \frac{24.45}{MW}$$

where,

C = chemical concentration in mg/kg
24.45 = the volume in liters of 1 mole of gas
at normal temperature and pressure
MW = the molecular weight of the chemical.

DENSITY OF LIQUIDS AND GASES

The density of a substance is defined as the mass of that substance per unit volume of the substance. The commonly encountered measures of density are: grams per cubic centimeter (g/cm^3), grams per liter (g/L), kilograms per liter (kg/L), pounds per cubic foot (lb/ft^3), and pounds per gallon (lb/gal).

To determine the density of a solid, liquid, or gas, determine the weight of a known volume and divide by the volume.

Other Useful Conversions - Water:

$$ppm = C \times \rho$$

where,

C = concentration (mg/L)
ρ = pollutant density (g/mL)

SIGNIFICANT FIGURES

Some uncertainties in measurement are inevitably introduced through factors such as human error, malfunction of measuring devices, and experimental bias (error in one direction, as by a ruler with "inches" that are too short). Information is useful only to the extent that one can be confident of its validity. To ensure such utility, each figure or digit in the numerical expression of a measurement should be significant. A significant figure may be defined as a number that is believed to be correct within some specified or implied limit of error. Thus, if the height of a man, expressed

in significant figures, is written as 5.78 feet, it is assumed that only the *last figure* may be in error. Clearly, any uncertainty in the first or second figure would remove all significance from the last figure (i.e., if you do not know the number of feet, it is useless to speak of inches). If we have reason to believe that the last figure will be in doubt by a specified amount, we may so indicate by expression such as 5.78 ± 0.01 feet.

To count the number of significant figures in a number, read the number from the left to the right and count all digits starting with the first digit **that is not zero**. The decimal point should be ignored because it is determined by the particular units employed, not by the precision of the measurement. Thus, the measurements 12.2 cm and 122 mm are equivalent, and both have three significant figures. Guard against introduction of uncertainty by arithmetical procedures. The following rules will be helpful.

Rule 1. In addition or subtraction, any figure in the answer is significant only if each number in the problem contributes a significant figure at that decimal level (that is, the level of greatest magnitude will determine how many significant figures should be carried in the answer):

$$
\begin{array}{r}
308.7812 \\
0.00034 \\
\underline{10.31} \\
319.09
\end{array}
$$

Rule 2. When a number is "rounded off" (nonsignificant figures discarded), the last significant figure is unchanged if the next figure is less than 5, and is increased by 1 if the next figure is 5 or more:

4.6349 → 4.635 (four significant figures)
4.6349 → 4.63 (three significant figures)
2.8150 → 2.82 (three significant figures)

Rule 3. In multiplication and division, the number of significant figures in the answer is the same as that in the quantity with the fewest significant figures:

$$\frac{3.0 \times 4297}{0.0721} = 1.8 \times 10^5$$

Rule 4. In a multi-step computation, it will be convenient first to determine the number of significant figures in the answer by rules 1-3 above, and to round off each number that contains excess significant figures to one or more significant figures than necessary. Then round off the answer to the correct number of significant figures. This procedure will preserve significance with minimum labor.

Example.

$$V = 4.3 \times \frac{311.8}{273.1} \times \frac{760}{784-2}$$

There are two significant figures in the number 4.3; therefore, the answer will have two significant figures. Round off according to rules 1 and 2, to one extra significant figure. Note that the presence of only one significant figure in the number 2 does not mean that there is only one significant figure in the answer because 784-2 = 782, which has three significant figures. Thus,

$$V = 4.3 \times \frac{312}{273} \times \frac{760}{782}$$

Solve and round off to two significant figures,

$$V = 4.8$$

APPENDIX L
SITE SUMMARY FORM

The following form should be included as an appendix to preliminary health assessments when available information is insufficient to perform a complete health assessment. A MultiMate version of this form may be obtained from ATSDR.

CERCLIS NO. _____

Date Prepared _____

Preparer _____

ATSDR SITE SUMMARY

I. GENERAL INFORMATION

Site Name: _____
(Include other names by which site is known.)

Region:_____ City:_____ County:_____ State:_____

Site Management Responsibility
[ ] Fund Lead   [ ] Enforcement Lead (PRP)
[ ] State Lead   [ ] Federal Facility

Remedial Schedule Status
[ ] PA/SI
[ ] Workplan Development
[ ] RI scheduled/under way
[ ] Other _____

II. DATA/INFORMATION REVIEW

(Review of EPA Site File(s) and, where appropriate, include State monitoring information)

II.A. Bibliography of Data/Information Sources:

Document	Date of Document
1_____	_____
2_____	_____
3_____	_____
4_____	_____
5_____	_____
6_____	_____

II.B. Brief Description of Site (include waste containment status)

_____

_____

_____

_____

_____

II.C. Previous ATSDR Recommendations   [ ] Yes   [ ] No

    [ ] Health Consultation (Verbal-Documented on SRC)
    Dates_____ _____ _____ _____ _____

    [ ] Health Consultation (Written Memo/Letter)
    Dates_____ _____ _____ _____ _____

    Explain Recommendations and Actions Taken: _____

_____

_____

_____

_____

_____

_____

_____

_____

_____

_____

_____

_____

II.D. Principal Contaminants of Concern

ON-SITE

Contaminant	Media	Ranges (specify units) Low	High	*Data Source and Date of Sampling
_____	_____	_____	_____	_____
_____	_____	_____	_____	_____
_____	_____	_____	_____	_____
_____	_____	_____	_____	_____
_____	_____	_____	_____	_____
_____	_____	_____	_____	_____
_____	_____	_____	_____	_____
_____	_____	_____	_____	_____
_____	_____	_____	_____	_____
_____	_____	_____	_____	_____
_____	_____	_____	_____	_____

OFF-SITE

Off-Site Data Reported [ ] Yes   [ ] No

Contaminant	Media	Ranges (specify units) Low	High	*Data Source and Date of Sampling
_____	_____	_____	_____	_____
_____	_____	_____	_____	_____
_____	_____	_____	_____	_____
_____	_____	_____	_____	_____
_____	_____	_____	_____	_____
_____	_____	_____	_____	_____
_____	_____	_____	_____	_____
_____	_____	_____	_____	_____
_____	_____	_____	_____	_____
_____	_____	_____	_____	_____
_____	_____	_____	_____	_____

*See Part II.A. for Appropriate Number

II.E. <u>Site Access Restrictions</u>

    1. [ ] Unrestricted Access
    2. [ ] Restricted Access (Explain Below)

    COMMENTS: (e.g., type of restrictions, restricting authority, etc.)

_____

_____

_____

_____

_____

II.F. <u>Removal Actions</u>

    1. Have removal actions occurred? [ ] Yes   [ ] No
    2. Describe the removal actions:

_____

_____

_____

_____

II.G. <u>Population</u>

    1. Distance to closest residence: ____

    2. Size of population within a ____ mile radius of the site: ____

    3. Special population concerns: [ ] Yes [ ] No
        (Are there schools, nursing homes, hospitals, parks, playgrounds, etc., within the radius?)

    COMMENTS:

_____

_____

_____

_____

II.H. <u>Environmental/Exposure Pathways</u>

II.H.1. <u>Groundwater</u>

Private Wells

a. There are private wells in use within the vicinity of the site.
    [ ] Yes   [ ] No   [ ] No data/information available within a radius of _____ miles.

b. Private well is used for:
    1. [ ] Drinking   4. [ ] Livestock
    2. [ ] Cooking   5. [ ] Irrigation of crops
    3. [ ] Other domestic uses 6. [ ] Other

c. There is reason to believe that the private wells are _____ are not _____ contaminated because of:

    1. [ ] Private well data
    2. [ ] Monitoring well data
    3. [ ] Public system data
    4. [ ] Other _____

d. The earliest documented date of private well contamination is:

_____

Public Wells

a. There are public/municipal wells in use within the vicinity of the site.
    [ ] Yes   [ ] No   [ ] No data/information available within a radius of _____ miles.

b. Public well water is used for:
    1. [ ] Drinking   4. [ ] Livestock
    2. [ ] Cooking   5. [ ] Irrigation of crops
    3. [ ] Other domestic uses 6. [ ] Other

c. There is reason to believe that the public wells are _____ are not _____ contaminated because of:

    1. [ ] Private well data
    2. [ ] Monitoring well data
    3. [ ] Public system data
    4. [ ] Other _____

d. The earliest documented date of well contamination is:

_____

Comments on private/public/irrigation well contamination:

_____

II.H.2. <u>Surface Water</u>
   a.   Are any of the following categories of surface water located
      on-site (or passing through the site):
      [ ] Drainage ditch (or intermittent stream)
      [ ] Stream or creek
      [ ] River
      [ ] Wetlands, pond, or lake

      Surface water is used for:
      [ ] Drinking   [ ] Cooking   [ ] Fishing
      [ ] Livestock   [ ] Swimming   [ ] Irrigation
      [ ] Other _____

      Surface water treated prior to use:
      [ ] unknown [ ] no [ ] yes
      Name of system owner: _____

   b.   Are any of the following categories of surface water adjacent
      to (bordering) the site:
      [ ] Drainage ditch (or intermittent stream)
      [ ] Stream or creek
      [ ] River
      [ ] Wetlands, pond, or lake

      Surface water is used for:
      [ ] Drinking   [ ] Cooking   [ ] Fishing
      [ ] Livestock   [ ] Swimming   [ ] Irrigation
      [ ] Other _____

      Surface water treated prior to use:
      [ ] unknown [ ] no [ ] yes
      Name of system owner: _____

   c.   Are any of the following categories of surface water impacted
      by the site:
      [ ] Drainage ditch (or intermittent stream): Distance to ___
      [ ] Stream or creek: Distance to ___
      [ ] River: Distance to ___
      [ ] Wetlands, pond, or lake: Distance to ___

      Surface water is used for:
      [ ] Drinking   [ ] Cooking   [ ] Fishing
      [ ] Livestock   [ ] Swimming   [ ] Irrigation
      [ ] Other _____

      Surface water treated prior to use:
      [ ] unknown [ ] no [ ] yes
      Name of system owner: _____

d. Summary of documentation of surface water contamination
    (include earliest date of contamination, discuss potential for
    contamination, discuss sampling that indicates surface waters
    may be contaminated):

    _____

    _____

    _____

    _____

    _____

    _____

    _____

    SOURCE(s): _____

II.H.3. <u>Soil</u>

   a. Off-site soil contamination confirmed: [ ] Yes   [ ] No
       Confirmed by: [ ] Sampling   [ ] Visible evidence

   b. On-site soil contamination confirmed: [ ] Yes   [ ] No
       Confirmed by: [ ] Sampling   [ ] Visible evidence

   c. The public is likely to come in contact with contaminated soil:
       [ ] Yes   Contact will occur: [ ] Off-site   [ ] On-site
                 Explain in Comments Section
       [ ] No

   d. On-site employees are likely to come in contact with
       contaminated soil: [ ] Yes   [ ] No

   e. The earliest <u>documented data</u> of soil contamination is:
       [ ] Off-site ____/____/____
       [ ] On-site ____/____/____

   f. Comments:

       _____

       _____

       _____

       SOURCE(s): _____

II.H.4. <u>Ambient Air</u>

    a. Release of volatiles or gases has been measured:
       [ ] Yes   [ ] No

Measurements were taken: [ ] On-site [ ] Off-site [ ] In Residence
    SOURCE(s): _____

There is a history of odor complaints in the vicinity of the site:
    [ ] Yes   [ ] No   Explain: _____

    _____

    _____

    SOURCE(s): _____

    b. A release of airborne particulates has occurred:
       [ ] Yes   Release confirmed by: [ ] Air sampling
                                        [ ] physical evidence
       [ ] No

    SOURCE(s): _____

    c. Comments on Ambient Air:

    _____

    _____

    _____

    SOURCE(s): _____

II.H.5. <u>Food Chain</u>

    a. <u>Crops</u>
    1. Are grown in the vicinity of the site: [ ] Yes   [ ] No
        Type [ ] Commercial agriculture [ ] Residential gardens

    2. Crops likely to be contaminated: [ ] Yes   [ ] No

    3. Verified by [ ] Sampling
               [ ] Observation (evidence of migration or stressed
                   vegetation)
    4. Crops (list) _____

COMMENTS: _____

_____

_____

b. Livestock/Domestic Fowl

1. Are kept in the vicinity of the site: [ ] Yes   [ ] No
    Type [ ] Commercial   [ ] Residential

2. Animals likely to be contaminated: [ ] Yes   [ ] No

3. Verified by [ ] Sampling
               [ ] Observed waste migration
               [ ] Reports of animal illness

4. Livestock/Fowl (list) _____

COMMENTS: _____

_____

_____

_____

SOURCE(s): _____

c. Fishing

1. Occurs in the vicinity of the site: [ ] Yes   [ ] No

2. Type: [ ] Commercial
            [ ] Recreational
            [ ] Food staple for area

3. Animals likely to be contaminated: [ ] Yes   [ ] No

3. Verified by [ ] Sampling
               [ ] Observed contamination
               [ ] Fish kills

d. Hunting

1. Is likely to occur in the vicinity of the site: [ ] Yes [ ] No

2. [ ] Game is sold commercial
    [ ] Recreational
    [ ] Game is a local food staple

3. Game is likely to be contaminated: [ ] Yes   [ ] No

4. Contamination verified by
    [ ] Sampling
    [ ] Observed contamination
    [ ] Reports of animal illness

5. Type of Game (list) _____

COMMENTS: _____

_____

_____

_____

SOURCE(s): _____

e. Comments of Food-Chain Contamination:

_____

_____

SOURCE(s): _____

III. REPORTED HEALTH COMPLAINTS

Reports of Increased Health Problem(s) Associated With the Site:

_____

_____

_____

_____

_____

_____

_____

_____

VI.   HUMAN EXPOSURE PATHWAYS

   A.   Opportunity for human exposure to groundwater contamination:
        1.   [ ] has occurred   [ ] is occurring   [ ] is not occurring
             [ ] is potentially occurring

        2.   If exposure occurred:
             [ ] >10 yrs ago   [ ] 1-10 yrs ago   [ ] <1 yr ago   [ ] unknown

        3.   Route of exposure:
             [ ] ingestion
             [ ] inhalation
             [ ] dermal contact

   B.   Opportunity for human exposure to surface water contamination:
        1.   [ ] has occurred   [ ] is occurring   [ ] is not occurring
             [ ] is potentially occurring

        2.   If exposure occurred:
             [ ] >10 yrs ago   [ ] 1-10 yrs ago   [ ] <1 yr ago   [ ] unknown

        3.   Route of exposure:
             [ ] ingestion
             [ ] inhalation
             [ ] dermal contact

   C.   Opportunity for human exposure to soil contamination:
        1.   [ ] has occurred   [ ] is occurring   [ ] is not occurring
             [ ] is potentially occurring

        2.   If exposure occurred:
             [ ] >10 yrs ago   [ ] 1-10 yrs ago   [ ] <1 yr ago   [ ] unknown

        3.   Route of exposure:
             [ ] ingestion
             [ ] inhalation
             [ ] dermal contact

   D.   Opportunity for human exposure to airborne contamination:
        1.   [ ] has occurred   [ ] is occurring   [ ] is not occurring
             [ ] is potentially occurring

        2.   If exposure occurred:
             [ ] >10 yrs ago   [ ] 1-10 yrs ago   [ ] <1 yr ago   [ ] unknown

        3.   Route of exposure:
             [ ] inhalation
             [ ] dermal contact

E.   <u>Opportunity for human exposure to food that has been contaminated through the food chain or by exposure to the site:</u>

  1. [ ] has occurred   [ ] is occurring   [ ] is not occurring [ ] is potentially occurring

  2. If exposure occurred:
    [ ] >10 yrs ago [ ] 1-10 yrs ago [ ] <1 yr ago [ ] unknown

  3. Route of exposure:
    [ ] ingestion

F.   Any other relevant human exposure information (historical exposure)?

_____

_____

_____

_____

_____

VII. <u>General Comments (optional):</u>

_____

_____

_____

_____

_____

_____

_____

_____

IV.　INTERVIEWS: PERSONS KNOWLEDGEABLE ABOUT THE SITE

The interview objectives are:
1.　to verify information found in the site file review and
2.　to acquire essential information not found in the site file(s).

A.　Name:_____ Organization_____ Date_____

Comments: _____

_____

_____

_____

_____

_____

_____

B.　Name:_____ Organization_____ Date_____

Comments: _____

_____

_____

_____

_____

_____

_____

C.　Name:_____ Organization_____ Date_____

Comments: _____

_____

_____

_____

_____

_____

D. Name:_____ Organization_____ Date_____

Comments: _____

_____

_____

_____

_____

_____

V.   **ATSDR SITE VISIT**

The purpose of the site visit is to verify information collected
during the site file review and the site interviews with
knowledgeable parties and to gather essential information not found
during the previous two steps.

Site Visits: [ ] Yes   [ ] No

              By Whom   Date

        _____   _____

        _____   _____

        _____   _____

        _____   _____

Comments: _____

_____

_____

absorption: the penetration of a substance into another.

accuracy: the nearness of a result or the mean of a set of results to the true or accepted value.

adsorption: the adherence of a gas, liquid, or dissolved substance to the surface of a solid.

aerosol: a suspension of fine liquid or solid particles in gas.

aerosolization: the dispersion of a liquid in the form of a fine mist.

aliphatic compounds: open-chain carbon compounds that are normally methane derivatives or fatty compounds.

alluvium: a general term for all sediment deposited in land environments by streams.

analyte: a chemical component of a sample to be determined or measured.

analytical method: defines the sample preparation and instrumentation procedures or steps that must be performed to estimate the quantity of analyte in a sample.

analytical spike: the addition of a known amount of a standard after digestion.

aquiclude: an impermeable stratum that acts as a barrier to the flow of groundwater.

aquifer: a permeable rock stratum below the earth's surface through which groundwater moves; generally capable of producing water for a well.

aquitard: a semipermeable formation that does not rapidly transmit fluids and hinders flow of groundwater.

aromatic compounds: compounds that contain a benzene ring.

artesian: refers to groundwater under sufficient pressure to rise above the aquifer containing it.

background correction: a technique to compensate for variable background contribution to the instrument signal and the determination of trace metals.

base-pair mutation: substitution mutation in which the wrong base is inserted into the DNA and is paired with its atural partner during replication resulting in a new pair of incorrect bases in the DNA.

bedrock: the continuous solid rock of the continental crust.

bioaccumulation: the process by which organisms retain chemical pollutants in their tissues at levels greater than in the ambient environment. This term is synonymous with bioconcentration.

biodegradation: the breaking down of a chemical compound into simpler chemical components under naturally occurring biological processes.

biomagnification: the process whereby chemicals concentrate to a higher level in organisms at one level in a food chain than in those at the preceding (lower) level in the food chain.

calibration: the establishment of an analytical curve based on the absorbance, emission intensity, or other measured characteristic of known standards. The calibration standards must be prepared using the same type of acid or concentration of acids as used in the sample preparation.

calibration blank: a volume of acidified de-ionized/distilled water.

carbonate: a mineral formed by the combination of the complex $(CO_3)_2$ with a positive ion; for example, $CaCO_3$ (calcite). Used for any rocks containing carbonate minerals, such as limestone and dolomite.

cation: a positively charged ion.

cation exchange: the reversible exchange between a cation in solution and another cation adsorbed onto any surface-active material, such as clay or organic matter.

cation exchange capacity (CEC): represents the extent to which the clay and humic fractions of the soil will retain charged species such as metal ions. The CEC is an important factor in evaluating transport of lead, cadmium, and other toxic metals. Soils with a high CEC will retain correspondingly high levels of these substances. The hazardous chemicals will be prevented from leaching into groundwater in the short-term, but in the long-term these soils may be a reservoir for continuing releases. Expressed in milliequivalents per 100 grams of soil (or clay).

chromosome aberration: changes in the number, shape, or structure of chromosomes.

chronic toxicity: a prolonged health effect that may not become evident until many years after exposure.

clastic: pertaining to a rock or sediment composed of broken fragments that are derived from preexisting rocks or minerals.

clay: (1) soil separate consisting of particles less than 0.002 mm in equivalent diameter. (2) soil material containing more than 40 percent clay, less than 45 percent sand, and less than 40 percent silt.

claypan: a compact, slowly permeable layer in the subsoil having a much higher clay content than the overlying material, from which it is separated by a sharply defined boundary. Claypans are usually hard when dry, and plastic and sticky when wet.

colluvium: any loose, poorly sorted mass of soil or rock material deposited by rapid, water-deficient processes, such as landslides, rockfalls, and mudflows; usually formed at the base of a steep slope; the soil or rock may range in size from clay to boulders.

comparability: a qualitative parameter expressing the confidence with which one data set can be compared with another. Sample data should be comparable with other measurement data for similar samples and sample conditions.

completeness: the percentage of measurements made that are judged to be valid measurements. The completeness goal is to generate sufficient amount of valid data based on project needs.

cone of depression: a conical depression in the water table immediately surrounding a well.

confining bed: includes terms "aquiclude," "aquitard," and "aquifuge" and is defined as a body of "impermeable" material stratigraphically adjacent to one or more aquifers.

contact: when a substance touches the body of a receptor.

continuing calibration: analytical standard run every 10 analytical samples or every 2 hours, whichever is more frequent, to verify the calibration of the analytical system.

control limits: a range within which specified measurement results must fall to be compliant. Control limits may be mandatory, requiring corrective action if exceeded, or advisory, requiring that noncompliant data be flagged.

correlation coefficient: a number (r) that indicates the degree of dependence between two variables (e.g., concentration and absorbency). The more dependent they are, the closer the value to one. Determined on the basis of the least-squares rule.

data quality objectives: qualitative and quantitative statements that specify the quality of the data required to support decisions during remedial response activities. Data quality objectives are determined based on the end uses of the data to be collected.

detection limit: the minimum concentrations that must be accurately and precisely measured by the laboratory and/or specified in the quality assurance plan.

dissolved metals: analyte elements that have not been digested before analysis and that will pass through a 0.45 μm filter.

DNA repair: repair of genetic material by cellular enzymes that can excise or recombine alterations in structure of DNA to restore original information.

degradation: a chemical reaction involving the breakdown of a molecule to form a simpler structure.

detritus: the accumulated particles of broken rock and skeletal remains of dead organisms.

dip (of a stratum): the angle in degrees between a horizontal plane and an inclined plane, measured down from horizontal in a plane perpendicular to the strike. Dip is measured with a clinometer.

discharge: the amount of water passing a given point in a given unit of time, as gallons per minute (gpm) or cubic feet per second (cfs).

dissociation: the separation of a chemical compound into simpler components.

divide: the line that separates adjacent drainage basins.

dolomite: a carbonate mineral, magnesium limestone $CaMg(CO_3)_2$; a rock composed mainly of dolomite is referred to as a dolomite rock or a dolostone.

dose: the amount of a contaminant that is absorbed or deposited in the body of an exposed organism for an increment of time. Total dose is the sum of doses received by a person from a contaminant in a given interval resulting from interaction with all environmental media that contain the contaminant. Units of dose and total dose (mass) are often converted to units of mass per volume of physiological fluid or mass of tissue.

drawdown: lowering of water level caused by pumping. It is measured for a given quantity of water pumped during a specific period or after the pumping level has become constant.

duplicates: identical splits of individual samples that are analyzed by the laboratory to test for method reproducibility. In this case, samples are split in the laboratory.

effluent: the discharge from a relatively self-contained source, such as from a sewage treatment plant or a nuclear power plant thermal discharge, generally carrying pollutants; the liquid substance, predominantly water, containing inorganic and organic molecules of those substances that do not precipitate by gravity.

equipment rinsates: the final analyte-free water rinse from equipment cleaning collected daily during a sampling event.

erosion: (1) the wearing away of the land surface by running water, wind, ice, or other geological agents, including such processes as gravitational creep. (2) detachment and movement of soil or rock by water, wind, ice, or gravity.

exposure (biology): an event that occurs when there is contact at a boundary between a human being and the environment with a contaminant of a specific concentration for an interval of time; the units of exposure are concentration multiplied by time.

exposure (geology): a place where solid rock is exposed at earth's surface.

field blanks: blanks are collected and analyzed to determine the level of contamination introduced into the sample because of sampling technique. They may consist of the source water used in decontamination and steam cleaning. At a minimum, one sample from each event and each source of water must be collected and analyzed.

field duplicates and splits: samples that have been divided into two or more portions while in the field. Each portion is then carried through the remaining steps in the measurement process. A sample may be replicated in the field or at different points in the analytical process. For field replicated samples, precision information would be gained on homogeneity, handling, shipping, storage, preparation, and analysis.

fluvial: of or pertaining to a river or rivers.

fold: a pronounced bend in layers of rock.

food chain: the transfer of food energy from the source (in plants) through a series of organisms that successively depend on each other for food.

frameshift mutation: mutation resulting from insertion or deletion of a base-pair from a triplet codon in the DNA; the insertion or deletion produces a scrambling of the DNA or a point mutation.

fume: solid particles generated by condensation from the gaseous state, generally after volatilization from a molten state. Formation is often accompanied by oxidation or other chemical reactions.

gas: formless fluids that can be changed to the liquid or solid state by increased pressure and decreased temperature.

gene mutation: a stable change in a single gene.

genetic toxicity: an adverse event resulting in damage to genetic material; damage may occur in exposed individuals or may be expressed in subsequent generations.

geologic map: a map showing the distribution, at the surface, of rocks of various kinds or of various ages.

geomorphology: the branch of geology dealing with the form and the general configuration of the Earth's surface and the changes that take place in the evolution of landforms.

grab sample: a discrete sample representative of a specific location at a given point in time.

groundwater: water beneath the surface of the ground in a saturated zone.

hydraulic conductivity: indicates the ease with which water will flow through the soil. It depends upon a variety of soil factors.

hydrologic cycle: the complete cycle of phenomena through which water passes from the atmosphere to the earth and back into the atmosphere.

hydrology: the science encompassing the behavior of water as it occurs in the atmosphere, on the land surface, and underground.

hydrolysis: (1) the formation of an acid and a base from a salt by interaction with water; it is caused by the ionic dissociation of water. (2) the decomposition of organic compounds by interaction with water, either in the cold or on heating, alone or in the presence of acids or alkalis.

hydrostatic pressure: the pressure of, or corresponding to, the weight of a column of water at rest.

infiltration: the movement of water into and through a soil.

instrument detection limit: the lowest concentration an analytical device is capable of measuring. This may be defined several ways: For example, (1) that concentration of analyte which produces an output signal twice the root mean square of the background noise may be determined under ideal conditions or (2) determined by multiplying by 3 times the standard deviation obtained for the analysis of a standard solution (each analyte in reagent water) at a concentration of 3 to 5 times the instrument detection limit on three nonconsecutive days with seven consecutive measurements per day.

internal standards: compounds added to every standard, blank, matrix spike, matrix spike duplicate, sample (for volatile), and sample extract (for semivolatile) at a known concentration before analysis. Internal standards are used as the basis for quantitation of the target compounds.

ion: an electrically charged particle of matter dissolved in water. For example, in water, salt forms sodium ions (Na) with positive charges, and chloride ions (Cl) with negative charges.

ion exchange: a reversible chemical reaction between a solid and a fluid mixture by means of which ions may be interchanged; used in water softening and separation of radioactive isotopes.

ionization: the process of converting a substance wholly or partially into ions.

laboratory control sample: a control sample of known composition. Aqueous and solid laboratory control samples are analyzed using the same sample preparation, reagents, and analytical methods employed for samples received.

laboratory quality assurance coordinator: an employee of a laboratory who has no analysis or production responsibilities and who implements QA and QC. This person is responsible for ensuring that all quality assurance problems are resolved.

latency: the period between stimulus application and response onset.

leachate: a solution obtained by leaching. Leachate from a sanitary landfill is a mineralized liquid with a high content of organic and inorganic substances. Any liquid, including any suspended components in the liquid, that has percolated through or drained from hazardous waste.

leaching: the continued removal, by water, of soluble matter from wastes, regolith, or bedrock.

limestone: a sedimentary rock consisting chiefly of the mineral calcite.

load (of a stream): the material carried at a given time, by a stream, by a current of water, by the wind, or by a glacier.

matrix: the predominant material comprising the sample to be analyzed. The most common matrices are water, soil/sediment, and sludge.

matrix spike: an aliquot of a matrix (water or soil) spiked with known quantities of compounds and subjected to the entire analytical procedure in order to indicate the appropriateness of the method for the matrix by measuring recovery.

matrix spike duplicate: a second aliquot of the same matrix as the matrix spike that is spiked in order to determine the precision of the method.

metabolic activation: the use of extracts of plant or animal tissue to provide enzymes that can convert a promutagen into an active mutagen, or a procarcinogen into an active carcinogen.

method blank: a blank sample run to ensure that reported analytical results are not the results of laboratory contamination.

method blank and spike: the distilled and/or deionized water for soil or sand spiked with known compounds or elements.

method detection limits: minimum concentrations of a substance that can be measured and reported with 99% confidence that the value is above zero. The sample is carried through the entire method under ideal conditions.

method of standard additions: the addition of three increments of a standard solution (spikes) to sample aliquots of the same size. Measurements are made on the original and after each addition. The slope, x-intercept, and y-intercept are determined by least-squares analysis. The analyte concentration is determined by the absolute value of the x-intercept. Ideally, the spike volume is low relative to the sample volume (~10% of the volume). Standard addition may counteract matrix effects; it will not counteract spectral effects. It is also referred to as standard addition.

minimal risk level (MRL): an estimate of daily exposure of a human being to a chemical (in mg/kg/day) that is likely to be without an appreciable risk of deleterious effects (noncarcinogenic) over a specified duration of exposure. MRLs are based on human and animal studies and are reported for acute ($\leq$ 14 days), intermediate (15-364 days), and chronic ($\geq$365 days). MRLs are published in ATSDR Toxicological Profiles for specific chemicals.

mutagenic: compounds with the ability to induce stable changes in genetic material.

organic carbon content: the amount of natural organic material in a soil has a strong effect on retention of organic pollutants. The greater the fraction by weight of organic carbon (foc), the greater the adsorption of organic chemicals. Soil foc ranges from less than 2 percent for many subsurface soils to more than 20 percent for a peat soil.

out of control: one or more of several conditions relating to the plotting of control data and indicating unacceptable results.

oxidation: the process of removing one or more electrons from an ion, atom, or molecule.

particulate: small, discrete, solid or liquid bodies, especially those suspended in a liquid or gaseous medium.

partitioning: the separation or division of a substance into two or more compartments. Environmental partitioning refers to the distribution of a chemical into environmental media (soil, air, water, and biota).

parts per million (ppm): a common basis of reporting water analysis. One part per million (ppm) equals 1 pound per million pounds of water; 17.1 equals one grain per U. S. gallon; 14.3 equals one grain per Imperial gallon.

percent solids: the proportion of solid in a soil sample determined by drying an aliquot of the sample.

perched groundwater: groundwater in a saturated zone separated from the main body of groundwater by unsaturated rock.

percolation: movement of contaminants from soil to groundwater occurring primarily by dissolution and transport with percolating soil water. Percolation is the volumetric flux per unit area of soil.

permeability: (1) the ease with which gases, liquids, or plant roots penetrate or pass through a bulk mass of soil or a layer of soil; varies with different soil layers. (2) the property of a porous medium relating to the ease with which gases, liquids, or other substances can pass through it; the capacity of rock or unconsolidated material to transmit a fluid.

pH: an expression of the acidity or alkalinity of a solution. It represents the minus base 10 logarithm of the concentration of free hydrogen ions. The range of possible pH values is 1 (most acidic) to 14 (most alkaline). The value of 7 represents neutrality.

pharmacokinetic: relating to the characteristic interactions of a drug and the body in terms of absorption, distribution, metabolism, and excretion.

photodegradation: the chemical breakdown of molecules caused by radiant energy.

plunge: the vertical angle between a fold axis and the horizontal plane.

porosity: the proportion, usually stated as a percentage, of the total volume of rock material or regolith that consists of pore space or voids; the volume percentage of the total soil volume not occupied by solid particles (i.e., the volume of the voids). In general, the greater the porosity, the more readily fluids may flow through the soil. An exception is clay soils, in which fluids are held tightly by capillary forces.

porous: containing pores, voids, or other openings that may or may not be interconnected.

potable: drinkable water.

precision: measure of the reproducibility of a set of replicate results among themselves or the agreement among repeat observations made under the same conditions.

preparation blank (reagent blank, method blank): an analytical control that contains distilled, deionized water and reagents, which is carried through the entire analytical procedure (digested and analyzed). An aqueous method blank is treated with the same reagents as a sample with a water matrix; a solid method blank is treated with the same reagents as a soil sample.

purge and trap: an analytical technique used to isolate volatile (purgable) organic compounds by stripping the compounds from water or soil by a stream of inert gas, trapping the compounds on a porous polymer trap, and thermally desorbing the trapped compounds onto the gas chromatographic column.

quality assurance: a planned system of activities (program) whose purpose is to provide assurance of the reliability and defensibility of the data.

quality control: a routine application of procedures for controlling the monitoring process. QC is the responsibility of all those performing hands-on operations in the field and in the laboratory.

reagent water: water in which an analyte is not observed at or above the minimum quantitation limit of the parameters of interest.

recovery: usually expressed as a percent. The numerical ratio of the amount of analyte measured by the laboratory method divided by the known amount of analyte added to the matrix (i.e., spiked sample) to be analyzed.

reference dose (RfD): an estimate (uncertainty spanning perhaps an order of magnitude) of a daily exposure (mg/kg/day) to the general public (including sensitive subgroups) that is likely to be without an appreciable risk of deleterious effects during a lifetime exposure (chronic Rfd) or exposure during a limited time interval (subchronic RfD).

regolith: the blanket, consisting of loose, non-cemented rock particles and mineral grains, that commonly overlies bedrock.

relief: the difference in altitude between the high and low parts of a land surface.

reporting detection limits: the same as method detection limits with consideration given for practical limitations, such as sample size, matrix interferences, and dilutions.

representativeness: expresses the degree to which sample data accurately and precisely represent a characteristic population, parameter variations at a sampling point, or an environmental condition. Representativeness is a qualitative parameter that is most concerned with the proper design of the sampling program.

sample holding times: times used to ascertain the validity of results based on the holding time of the sample from time of collection to time of analysis or sample preparation. Holding times may vary depending on the analysis, EPA regional preferences, and other factors.

saturated zone: that part of a water-bearing material in which all voids, large and small, are filled with water.

scale (of a map): the proportion between a unit of distance on a map and the unit it represents on the Earth's surface.

screening: the process of rapidly identifying potentially important chemical contaminants and exposure pathways by eliminating those of known lesser significance.

seepage: (1) the appearance and disappearance of water at the ground surface; (2) the type of movement of water in unsaturated material; distinguished from percolation, the predominant type of movement of water in saturated material.

semivolatile compounds: compounds amenable to analysis by extraction of the sample with an organic solvent. Used synonymously with base neutral acid or extractable compounds.

serial dilution: the dilution of a sample by a known factor. When corrected by the dilution factor, the diluted sample must agree with the original undiluted sample within specified limits. Serial dilution may reflect the influence of interferents.

silt: a fine-grained sediment having a particle size intermediate between that of fine sand and clay (between 0.02 and 0.002 mm in diameter).

sink: a large solution cavity open to the sky, generally created by collapse of a cavern roof.

sludge: any solid, semi-solid, or liquid waste generated from a municipal, commercial, or industrial waste water treatment plant, water supply treatment plant, or air pollution control facility.

solids — total, dissolved and suspended: (1) Total solids represent the sum of dissolved and suspended solids; (2) dissolved solids are in true solution and cannot be removed by filtration. Their origin lies in the solvent action of the water in contact with the earth's minerals; (3) suspended solids are those not in true solution and removable by filtration, and they may be imparted from small particles of insoluble matter, from turbulent action of water on soil, or from domestic and industrial wastes.

solution cavities: numerous holes and openings that form in carbonate rocks as a result of chemical activity.

sorption: the act of taking up and holding a chemical or substance by either adsorption or absorption.

spikes: known amounts of specific chemical constituents added by the laboratory to selected samples to test the appropriateness and recover efficiencies of specific analytical methods within the actual sample matrices.

standard deviation: the square root of the variance of a set of values.

stratigraphy: the definition and description of major and minor natural divisions in layered rocks, such as groups, formations, and members.

surrogates: compounds added to every blank, sample, matrix spike, matrix spike duplicate, and standard and used to evaluate analytical efficiency of the method by measuring recovery. Surrogates are brominated, fluorinated, or isotopically labelled compounds not expected to be detected in environmental media. These are used typically in organic methods.

tentatively identified compounds (TICs): compounds detected in samples that are not target compounds, internal standards, or surrogate standards. Up to 30 peaks (those greater than 10% of peak areas or heights of nearest internal standards) are subjected to mass spectral library searches for tentative identification.

total metals: analyte elements that have been digested before analysis.

transformation: the act of changing in composition or structure.

vapor: a substance in the gaseous state as distinguished from the liquid or solid state; volatile form of substances normally in the liquid or solid state at normal temperature and pressure.

vapor pressure: the pressure exerted by a vapor, either by itself or in a mixture of gases; often taken to mean saturated vapor pressure, which is the pressure of a vapor in contact with its liquid form.

variance: the sum of the squares of the difference between the individual values of a set and the arithmetic mean of the set, divided by one less than the number of values.

volatile: characterized by rapid evaporation.

volatile compounds: compounds amenable to analysis by the purge and trap techniques. Used synonymously with purgable compounds.

volatility: the ability of a chemical to vaporize or evaporate.

volatilization: the conversion of a liquid or solid into vapors.

volatilize: to evaporate or turn into vapor from a liquid or solid state.

washout: occurs when falling raindrops or snowflakes collide with and retain large aerosol particles; it is effective only in the removal of large particles.

watershed: the area, defined by physical drainage divides, that is drained by a stream or stream system.

water table: the upper surface of the zone of saturation where the water pressure is equal to atmospheric pressure.

M.1. REFERENCES

Landau SI, editor-in-chief. *International dictionary of* medicine and biology. New York: John Wiley & Sons, 1986.

Brown, KD, Evans GB Jr, Fentrup, BD *Hazardous waste land* treatment. Boston: Butterworth Publishers, 1983.

Mish FC, editor-in-chief. *Webster's Ninth New Collegiate Dictionary. Springfield, Massachusetts: Merriam-Webster, Inc., 1987.*